Panda Nation

Panda Nation

The Construction and Conservation of China's Modern Icon

E. ELENA SONGSTER

OXFORD
UNIVERSITY PRESS

Oxford University Press is a department of the University of Oxford. It furthers the University's objective of excellence in research, scholarship, and education by publishing worldwide. Oxford is a registered trade mark of Oxford University Press in the UK and certain other countries.

Published in the United States of America by Oxford University Press
198 Madison Avenue, New York, NY 10016, United States of America.

First issued as an Oxford University Press paperback, 2020

Library of Congress Cataloging-in-Publication Data
Names: Songster, E. Elena, author.
Title: Panda nation : the construction and conservation of
China's modern icon / E. Elena Songster.
Description: New York, NY : Oxford University Press, [2018] |
Includes bibliographical references and index.
Identifiers: LCCN 2017052752 (print) | LCCN 2018001326 (ebook) |
ISBN 978-0-19-939368-8 (Updf) | ISBN 978-0-19-939369-5 (Epub) |
ISBN 978-0-19-939367-1 (hardcover : alk. paper) |
ISBN 978-0-19-753357-4 (paperback : alk. paper)
Subjects: LCSH: Pandas—China—Symbolic aspects. | Pandas—China—
Political aspects. | Pandas—China—Conservation.
Classification: LCC QL737.C27 (ebook) | LCC QL737.C27 S6394 2018 (print) |
DDC 599.7890951—dc23
LC record available at https://lccn.loc.gov/2017052752

To Matt and our wonderful children, Dylan, Clara, and Liam

Contents

Acknowledgments

AS THIS PROJECT progressed and extended, it involved more people and accumulated more debt in gratitude. Many acknowledgments end with appreciations to the family; I will begin there as they are the people most directly affected by this project and my work on it, good, bad, and otherwise. I could not ask for a better partner in life than my husband Matt Diffley. I am especially appreciative of his focus on our family and our three wonderful children, Dylan, Clara, and Liam. These kids have brought us true joy. And each in his and her own way has reminded me of the broader world and shown me things I would only be able to see with their assistance and individual perspectives. Matt's incredible support for my work and career manifests itself in a multitude of ways and has pulled him to many places, from Sichuan to San Diego, and from OK-City to Oaktown, even while navigating his own career in medicine.

The person next closest to this project in recent years is Susan Ferber. Her excellent editing, guidance, and shepherding have all made this a better book, and her persistent faith in this project and periodic panda cartoons helped me see it through to the end.

The giant panda's home is in China. It was there that this project bloomed, and it is to the people in China who became part of this story that I am most indebted. It was a true honor to work with Pan Wenshi. I not only learned a tremendous amount from his broad knowledge and experiences, but also from watching him make things happen around him. In addition to becoming a central part of this story, he facilitated my work with introductions, insights, and an intrepid spirit. I am particularly indebted to Zhong Zhaomin for generously sharing his time and story. He continues to be an inspiration to me. Thank you to Wang Dajun for giving expert guidance to and through Wanglang and for introducing me to the people who work there, to Lü Zhi for providing the bookends for my research with insightful interviews and continuing to be a patient consultant. I am deeply grateful to all of those who offered time to share their personal histories in interviews and conversations. Wang Menghu, Qing Jianhua, Jin Jianming, Hu Tieqing, Feng Yunwu, and Feng Wenhe all generously shared their time and perspectives. A tremendous thank you to Chen Youping and Jiang Shiwei for accommodating me so hospitably at Wanglang, sharing their experiences with me, and for their particular kindness and generosity on my return trip many years later. I am also very grateful to Hu Jinchu for his prolific publications, important historical role, and for kindly sharing his knowledge. Thanks also to Zhu Xiaojian and Long Yu for thoroughly responding to my many inquiries, to Wang Hao for bringing the project back to its origins, to Binbin Li for continuing the narrative, and to Yin Lijie, Yuan Yan, and Juan Li for all and various kinds of help. I would like to thank Wen Bo for sharing his environmental enthusiasm and insight. Thank you also to Mabel Lam and Huang Shiqiang for sharing their time, history, and providing many introductions.

The Feng and Li families offered me a home away from home in China, complete with enthusiastic support for my research and constant nourishment for my mind and belly. I am particularly grateful to Feng Binbin for introducing me to key people in Chengdu and am still seeking ways to properly honor her memory. Thank you to Zuo Ye for his poetic personality and to Feng Bingkun and Nan Yu and the rest of their family for their interest and support and for so graciously extending their family to us.

Even while the book in many ways has taken on a life of its own, its origins have solid roots in the UC San Diego graduate program. Joseph W. Esherick opened my eyes to the potential the panda had to offer and Paul G. Pickowicz demonstrated through his own participation in the narrative the expanse of the panda's reach and its relevance. As I have advanced through my career, I find myself repeatedly thankful for their training, guidance, and support as co-advisors. Their exemplary devotion to their students enlightened all of us to the value of community. Each fellow graduate student I thank here is an additional expression of gratitude to our advisors for the collaborative culture that they fostered. Many sources in my stacks have Sigrid Schmalzer's handwriting scribbled on the top, as she set them aside while pursuing *yeren*; source sharing was but one of the many ways she offered support for this project. Sue Fernsebner, Cecily McCaffrey, Liu Lu, Joshua Goldstein, Michael Chang, Gerry Iguchi, Chris Hess, Jeremey Brown, Charles Musgrove, James A. Cook, Miriam Gross, Xiao Zhiwei, Zheng Xiaowei, and our good-natured librarian Richard

Wang each touched the project in one way or other with suggestions, questions, encouragement, critiques, or comradery. Daphon Ho, Ellen Huang, and Matt Johnson offered insights that made it into my footnotes. Andrew Morris and Elya Jun Zhang subsequently created space for panda programming. Gabriela Soto Laveaga, Angela Vergara, Adam Warren, Sarah Malena, and Donald Wallace bridged historical regions and rounded my understanding.

Beyond this cohort, Ye Wa, Kim Kono, and Yun-Chiahn Chen brought important nuance to this narrative. Angela Borda gave me crucial and good-natured editing advice. Suzanne Thomas and Lisa Tran have been voices of reason from research to the real world. Michael Hathaway, Janet Sturgeon, and Chiho Sawada offered important feedback and became valued colleagues in the fields of Asian environmental studies. Henry Nicholls joined in pursuit of panda history and pointed interested readers in my direction. Over the years I also had special assistance from students in research, editing, and proofreading from Stephanie Tso, Allen Wang, and Andrew Wozniak.

Robert Marks, Marta Hanson, Naomi Oreskes, Mark Hineline, and Martha Lampland broadened and deepened my lines of inquiry. Suzanne Cahill, Stefan Tanaka, and Tak Fujitani offered insights long ago that appear indirectly in this and my other work. Wen-hsin Yeh was my first mentor; she offered me early encouragement and continues to inspire me.

Several institutions and organizations have funded various aspects of this project: University of California San Diego, J. William Fulbright Foundation, University of Oklahoma's College of Arts and Sciences Faculty Enrichment Grant and Junior Faculty Research Fellowship, Faculty Fellowship Program for Scholarly Exchange in Big XII Universities, University of San Francisco's Kiriyama Research Fellowship, Saint Mary's College of California's Faculty Development Fund, Frank J. Filippi Endowment, and dedicated sabbatical time all supported the research, writing, and editing of this book.

I must also thank the people at these institutions, especially my colleagues in the History Departments at University of Oklahoma and Saint Mary's College of California. At OU so many more deserve individual mention, the list is long so know that I treasured my time there with specific thanks to Robert Griswold who was a particularly supportive chair, Elyssa Faison who offered important insight about Japan, Donald Pisani for his mentoring, Cathy Kelly for the title, and Sandie Holguin, Judy Lewis, and Melissa Stockdale whom I often sought out for advice.

At SMC, I want to thank all my colleagues in History: Myrna Santiago, Paul Flemer, Carl Guarneri, Brother Charles Hilken, Gretchen Lemke-Santangelo, and Aeleah Soine for tremendous support all around, Kathy Roper for translation assistance, and Jennifer Heung and my many other wonderful colleagues across the College. Even with all of this support and assistance the book remains imperfect and all omissions and flaws are my sole responsibility.

Before and beyond academics, a special thanks to my friends Em Dore and Cyndi Wong and my ever-extending family, especially Robert Diffley who offered useful feedback and, with Brenda's help, was instrumental in aiding the completion of this book by offering loving care for his grandchildren. To Erin Diffley and Melinda Love who, as exuberant aunties, honor the memory of Laurie in the love they shower on her grandchildren. Thank you to Peter Songster, my brother, for introducing me to the allure of history when we were adolescents and to Catie, Joshua, Benjamin, and Isaac. Finally, I feel deep gratitude toward my parents, Donald and Giuliana Songster, whose lifelong encouragement has made all the difference for me, Matt, and the kids. They brought me here and are lovingly always there.

Panda Nation

Introduction

THE GIANT PANDA is the most ubiquitous rare animal on Earth. Outside of China, the giant panda is a logo for a fast food chain, a software program, a licorice company, and a nonprofit environmental organization, among others. Within the People's Republic of China (PRC), the exclusive home of wild giant pandas, it is a logo on just as wide a spectrum of products and causes; it can be found on items from chopsticks and toilet paper to cell phones and automobiles, and it was memorably a mascot for the 2008 Olympic Games. Artists in China incorporate it into designs on porcelain vases, silk brocade, and traditional-style brush painting. Although the panda's image pervades China and the world, the scarcity of the actual animal and its shrinking habitat has been the source of focused concern among scientists and the public alike.

The panda is broadly known and widely recognized, but surprisingly it cannot claim deep historical prominence. Even though the panda has lived in the territory of present-day China for millions of years, its popularity within China is largely a post-1949 phenomenon. The panda's lack of established historical significance is contrary to popular expectations and a key reason why it is historically interesting. The transformation of the giant panda from its unsung and hidden existence in the remote mountains of western China to a position of national and international recognition and influence is in many ways a byproduct of a larger story: the rise of the People's Republic of China.

The history of the giant panda's emergence as an icon offers a unique lens onto China's government, its people, and its natural environment. It illuminates the significance of territory and place, the integration of science and government, and the emotional pride of a nation and its people as fundamental to the composition of China and its history. The PRC has astutely presented the giant panda as a means to honor select member states of the international community, offering the world a friendly face of China. Some have argued that China hides its true self behind this adorable ambassador. If one studies this animal and its image closely, however, it becomes apparent that the panda reveals much more about China than it masks. The giant panda represents not simply the country of China, but also the many nuanced complexities of its nationhood.

The emergence of the giant panda as a national icon was made possible in part by its own striking natural appearance and allure, but ultimately was the result of China's effort to define itself as a nation. As the subject of government-directed science and popular nationalism, the giant panda's rise went in tandem with the dramatic ascent of China to a position of broad global influence. Upon the founding of the PRC, the young government struggled to create a nation out of a country that had been shattered by decades of international and civil war. In this effort, the state linked people to their home territory through myriad steps and missteps that both inspired fervent nationalistic spirit and brought about catastrophic failures. While this book is not intended as a comprehensive history of the PRC, it aims to demonstrate the significance of the giant panda in integrating the development of science, China's natural environment, and widespread nationalism in China's evolution and identity.

When this project began, the giant panda operated as a means to examine the history of modern China. It quickly ceased being simply a means to an end, however, and became an end itself. As such, the book differs from other books on the history of the PRC and other books about the giant panda. Following the panda's story through modern Chinese history highlights unexpected intersections of humanity and awe-inspiring wilderness, hope, and disillusionment. And China is at every turn transforming in ways that look familiar but alter expectations. This book underscores the importance of nature in China's twentieth-century history, despite the country's much discussed reputation for damaging nature, from deforestation to pollution. It demonstrates the initiative of China's scientists and policymakers in engaging with China's nature without external interests and interference. A bridge to nature, the panda also integrated the urban centers with local officials and ethnic minority villagers in China's remote regions. As a point of pride, the panda symbolized cultural and economic shifts before it was used as a diplomatic tool. As a national subject, it helped reintegrate China with the world.

There are many extensive scientific studies of the giant panda and numerous articles by researchers from across the globe that have contributed further to the understanding of this animal's unique anatomy, ecological habitat, and genetics.[1] *Panda Nation* examines those who studied the panda and the political and historical parameters in which they worked as participants in the history that surrounded them. A number of works have offered cultural contextualization for the panda fever that spread across the globe. In the early twentieth century Ruth Harkness, the first person to bring a live panda to the west, wrote her own story, which has since been frequently retold, of extraction and transforming the world's impression of pandas. Others have charted the impact of the beloved animal that Harkness introduced and chronicled the journeys of other pandas in the west.[2] These books are crucial to better understanding just how and why the giant panda has become an international phenomenon. The important story of the panda's relationship to China, however, is not told in these global studies.

In order to understand how the giant panda became a national icon in China, this book draws on Chinese- and English-language sources from local, provincial, and national archives, newspapers, scientific articles, government policies, surveys, books, laws, and insights from experts, officials, and zoo keepers. It analyzes artwork, advertisements, maps, photographs, and statistical data, as well as first-hand observations of the pandas and the habitat itself. It is a reflection of a truism, stated by US environmental historian Mark Hineline, "environmental history must be studied just as the environment is experienced, through all of the senses."[3]

Panda Nation contributes to a growing compendium of environmental histories and environmental studies on China and Asia, more broadly speaking, once the purview of historical geographers and economic historians. Recognition of the importance of the environment to understanding China's history and society has gained tremendous traction among scholars in recent years. At the same time the field of animal studies has underscored that there is much to be learned about humans by examining their relationship with animals.[4] *Panda Nation* keeps the panda as a central focus of the book at the same time that it examines the historical and political context of the animal's emergence from obscurity to prominence.

This study demonstrates that the giant panda is not just another animal. Not only is the panda biologically unique in the animal kingdom, but in the human world, its allure and political power are unmatched.

Powerful countries that have asserted historical importance on an animal have chosen imposing and majestic animal symbols. There are other cute symbolic animals, but none that have been as able to command as much diplomatic sway and political force. The giant panda's emergence as a national icon was possible precisely because it happened as an integral component of China's transformation as a nation. It was not contrived; it was a coevolution, both an outgrowth of and a contributor to China's identity.

The rise of the giant panda to its current position of global familiarity, as both an animal and a symbol, began with scientific curiosity and evolved in multiple directions at once. It transformed from a local oddity to a global curiosity when it stepped out of the mountains and onto the international stage in 1869 as a new entry in the growing western catalogue of world species. Since that time, the panda has inspired unwavering interest from scientific communities across the globe. The integration and mastery of western science have been central to discourse about China's national identity from late nineteenth-century self-strengthening debates to current advances in space technology, including the 2016 completion of world's largest radio telescope.[5] Panda-related scientific studies appear in an expanding diversity of research fields, both reflecting and advancing China's position in international scientific exchange. During the early twentieth century, paleontological studies expanded taxonomical debates about the panda.[6] Scientific research on the giant panda demonstrates how the Chinese government integrated science into state building, and how the scientific approaches that the state favored varied markedly from decade to decade. Scientific papers range on subjects from histology and digestive parasite analyses to captive breeding. In fact, the giant panda expanded the fields of science pursued in China from the introduction of conservation biology to the panda genome project.[7] When scientists began examining the behavior of wild populations, they drew the focus of the nation to its home range in China's western mountain hinterlands. As a research subject, the wild panda population forced China to reckon with the potential of nature and the limits of China's natural resources. As an important subject of scientific study, the giant panda illuminates the centrality of science to China's development during the twentieth century, particularly under the PRC government. Indeed, a growing literature on the history of science in China testifies to the expanding recognition that science has been one of the central pillars supporting the structure and identity of China. Moreover, as historian of science Fa-ti Fan has pointed out, the history of twentieth-century science in China highlights the growing need to think about different ways of viewing modernization and viewing scientific developments from a more transnational perspective.[8]

Curiosity about the secrets, substance, and potential in nature as well as experts' hubris about the ability of science to improve and fix nature have shaped the evolving relationship between modern China's people and their natural environment. As central as science has been to the construction of the modern Chinese state, the territory in which China exists has been just as important a defining force. China's natural environment reflects the nation's expansive self-perception and simultaneously is a reminder of the limits of nature's ability to endure such expansion. The ways that China's government, scientists, and citizens perceived and treated the giant panda over the decades reflect the changing ways that each of these groups viewed nature over time. The panda not only was subject to these shifts, but also helped to create them.

One of the reasons the panda was able to influence perspectives of nature in Chinese society was because it was a source of national pride. The nation embraced the animal, fueling nationalistic interest in and concern for the giant panda. It became an expression of nationalism, a tool for diplomacy, and a means for international cooperation and scientific exchange. China's relationship with other nations and its relative position in the world is punctuated with panda politics.

Panda Nation traces the emergence of the giant panda from a virtually unknown animal to a national treasure from the 1950s to the present day. Because the panda was almost entirely absent from China's cultural past, the general populace did not take a strong interest in it until after 1949, as Chapter 1 shows. An upsurge of scientific studies on the giant panda's numerous curiosities aroused both domestic and international interest. Because of its biological peculiarities, the panda became a useful tool for the new communist government's agenda in mass education. Although the panda was first introduced to the public through science, its attractive and unique appearance inspired popular adoration. From the 1950s, when the panda first appeared in the Beijing Zoo, the Chinese people were increasingly intrigued and within the decade had embraced it as the new member of the national family.

By examining nature protection policies from the 1950s and 1960s, Chapter 2 demonstrates a shift in perceptions of nature as either a hurdle to overcome or a wellspring of resources and hidden knowledge. This chapter examines how the Great Leap Forward (1958–1959) and subsequent famine forced the government to reassess the relationship between the state and its natural environment. Government officials acquired a new perspective on nature and for the first time saw it simultaneously as a means of generating revenue and a place in need of protection. When these new policies are seen in the context of broader communist ideological rhetoric, they illustrate that government officials and scientists engaged in vigorous dialogue and expressed varied views on the place of nature in communist thought. Not dissimilar to arguments posed by Progressive-era American conservationists, the rhetoric of PRC policies about nature protection during the 1950s emphasized the importance of resource protection for the sake of state development.

Chapter 3 goes to the local level to investigate how the national nature protection policy was implemented on the ground. The Wanglang Nature Reserve became the first space demarcated as a panda reserve and thus the first experiment in engaging local people in the national cause of panda protection. Local bureaucrats, villagers of the Baima ethnic group, and scientists grappled over their differing interpretations of the panda's elevated status as a protected species. The interactive nature of their efforts to define China's natural environment hightlights the expansion of the panda's role in society during the 1960s.

Many advances in nature protection during the early 1960s were suddenly halted with the onset of China's Cultural Revolution (1966–1976). However, activities at this reserve continued. The Wanglang Nature Reserve hosted representatives from China's top scientific institutes on the first species-specific giant panda survey in 1967. Chapter 4 offers a counter-narrative to the characterization of the Cultural Revolution as a decade of chaos. During this era, the government endorsed many scientific endeavors designed to highlight the glory of China's nature and assert its continued scientific prowess. In addition to being an approved scientific subject, the giant panda became a popular expression of nationalism during this era. Because the panda was seemingly apolitical and benign—even cute—it could never make any political gaffes. It was repeatedly reproduced as a politically appropriate image and demonstrated impressive durability amidst changeable political winds.

One of the most salient examples of the giant panda as a national symbol, the offering of state-gift pandas to other countries, became an important dynamic in China's shift away from the isolation demanded by the Cultural Revolutionary call for self-reliance. State-gift pandas were among the most successful efforts by China to forge a new, softer, and more approachable image as it strove for greater international recognition and integration. Chapter 5 illustrates how these high-profile gifts had a profound effect on the wild panda population as well as international politics. The impact of "panda diplomacy" on China's wild pandas inspired new protection policies during the 1970s. This chapter also discusses the continued relevance of the wild population to the historical, cultural, and political implications of panda gifts.

Giant pandas markedly aided China's efforts to reach out to the global community. Internationalization became one of the hallmarks of the Deng Xiaoping era (1978–1989). Chapter 6 examines the ways that the world began to make a deeper imprint on China under Deng's leadership, even reaching giant panda territory in the remote mountains of western China. The transition in leadership from Mao Zedong to Deng Xiaoping is the context for two panda-starvation scares juxtaposed in this chapter. The responses to these scares demonstrate a dramatic shift in both the government and the people's perceptions of nature. Nature no longer simply belonged to the state. Instead, efforts to save the panda transformed nature and its protection into the common responsibility of the entire population. This mirrored the way the country was beginning to privatize in both big and small ways.

Chapter 7 examines the ways that the giant panda continued to enhance China's integration with the international community from the 1990s on. This era transformed panda country. The Wanglang reserve became a site where the local Baima villagers, scientists, NGOs, and tourists from both China and abroad cooperated and competed to redefine the giant panda's role in its home range. The colorful ethnic presentation of the Baima people initially proved to be an asset to World Wide Fund for Nature (WWF; formerly the World Wildlife Fund) efforts to instigate ecotourism as an environmentally friendly way for the local people to subsist. This new lucrative industry, however, took on an identity independent of panda preservation efforts. Advocates for panda preservation have since returned their focus to early uses of the reserve as a scientific research station for the development of conservation strategies.

In part as a response to the starvation scare of the 1980s, the Chinese government decided to cease giving pandas to other nations and instead loan them out for short-term overseas exhibits. Instead of permanently depleting the wild population with panda gifts to other nations, a panda on loan could visit multiple countries and remain the property of China. These short-term loans proved to be an effective means to generate income. The practice reflected a China that was now permitted to profit financially. China came under fire, however, for profit seeking at the expense of pandas' health in spite of the fact that these loans theoretically were originally introduced as a means to better protect the depleted species. The panda once again became emblematic, this time of the astounding economic transformation of China and the permeating forces of the market. The controversy over the short-term loans also required a reconfiguring of panda-loan policy.

The last chapter traces the metamorphosis of the phenomenon of short-term loans into scientific loans, as well as the perpetual effort to denounce the presence of politics in the selection of recipient countries and institutions. Politics, however, did not fade with the introduction of the market to panda diplomacy. On the contrary, the complexity of the giant panda's political role peaked when the PRC offered a pair of pandas to the island of Taiwan in 2005. This gesture ignited more than three years of political controversy and nationalistic debate about the meaning and political implications of the gift. This new era of panda loans encapsulated the political repositioning of China as it gained strength through economic growth. At the same time, the relationship between China's wild and captive panda populations grows increasingly complicated as more international zoos are added to the list of scientific research partners and China pushes forward with its captive-to-wild reintroduction agenda.

The basic paradoxes that characterized the giant panda upon its introduction to western science in the late nineteenth century have only multiplied as the panda has grown in cultural importance, domestically and globally. Once fascinated by the concept of a vegetarian carnivore, the world now contemplates the captive baby boom of a threatened species. The panda woos visitors to zoos with its primal charm and is simultaneously the subject of some of the world's most advanced technologies available for probing its scientific mysteries. As its habitat in the uplands of China's western mountains continues to shrink, the panda's value as a species and its prominence in global culture continue to grow. In the end, the importance of the giant panda in China and throughout the world is dependent upon the people who surround it and the meaning they imbue it with as they continue to forge their national identity.

1

Obscurity, Oddity, and Icon

THE EMERGENCE OF THE GIANT PANDA AS ANIMAL AND IMAGE

THREE LARGE, RED scrolls unfurled to the ground. Emblazoned on the first scroll were the Chinese characters "华 美." On the second was the romanization of these characters, "Hua Mei." The third scroll showed their English translation: "China USA." Hua, a character used to denote China, also means "magnificent." Mei is an abbreviation of "America" and also means beauty. The crowd of representatives from both China and the United States watching the scrolls unfurl applauded. They had gathered on December 1, 1999, for the "unveiling" of the name chosen for the new giant panda cub born a few months earlier at the San Diego Zoo in California. Members of the State Forestry Administration of China chose this name to celebrate the collaborative effort between the two countries that made possible the birth of the first giant panda to be successfully conceived and reared in the United States.

When Hua Mei was born, she represented almost every issue that surrounds the giant panda today. As the product of an international breeding experiment, she was a symbol of international scientific cooperation. As the offspring of a wild-born male and captive-born female, she literally embodied elements from the two habitats that comprise the main venues for panda research and panda existence. As a captive-born panda herself, Hua Mei lives in the human realm, visible, accessible, and researchable. However, she exists ostensibly for the sake of her wild counterparts and their habitat. The effects of this type of international cooperation have spread beyond zoos, breeding centers, and laboratories to China's mountains and the wild panda population. In the past, pandas were displayed outside of mainland China after being sold to zoos, given to foreign governments, or placed on short-term loans. Now, with the exception of the pandas sent to Taiwan in 2008, the animals reach overseas destinations through long-term scientific loans that entail substantial annual fees and research exchange agreements between China and the host zoo. Money generated from these loans is designated for wild-panda protection and research projects in China.[1] Additional money is usually paid to China when a giant panda cub is born in a foreign zoo.

While Hua Mei was at the San Diego Zoo, the informational signs about wild pandas, reproductive science, and China that lined her enclosure gave her three distinct identities: a representative specimen of captive breeding science, a wild endangered species, and a symbol of the distant People's Republic of China. Yet, even as she was portrayed as a Chinese diplomat, Hua Mei also became a popular representative of San Diego, featured in airport kiosks and on postcards, billboards, and festive streetlight flags. As such, she became the showpiece of the zoo and a symbol of her host city. Under the terms of the panda loan agreement, however, she remained the property of China. In this way, China and endangered species advocates in the United States could ensure the continued channeling of zoo revenue into wild-panda protection.[2] Asserting ownership over all giant panda offspring enabled China to protect its possession of giant panda progeny and maintain an ongoing connection between the giant panda and its homeland.[3]

Bestowing pandas on foreign countries as state gifts and later sending them to zoos on scientific research exchanges have been among the most conspicuous expressions of the panda as a symbol of Chinese nationhood. The panda only rose to this position of cultural prominence in China, however, during the latter half of the twentieth century. Although scientists had placed a great deal of emphasis on the panda's prehistoric ties to the territory of present-day China, the giant panda's political potency is grounded in the very fact that its cultural significance within the PRC is effectively severed from China's imperial past. The giant panda's lack of traditional symbolism became an attractive attribute of the animal during the Mao era.

Today the strong association between the image of the giant panda and the nation of China is reinforced through various art forms. The giant panda is a common motif in countless "traditional" arts and crafts. The presence of the panda in various art forms attaches a sense of tradition, history, and cultural value to the popular image of the giant panda, both within and outside China. The giant panda, however, only began appearing in "traditional" art between forty and fifty years ago. Unlike the dragon, which represents the emperor and his power, the giant panda is not an animal of prominent ancient symbolic importance in China. It is not found on ancient metallurgy or pottery, nor does it have a conspicuous presence in Tang poetry or Song paintings, Yuan plays, Ming novellas, or Qing porcelain. While the phoenix, the tiger, the crane, and even the cicada and cricket hold prominent places in Chinese cultural history, the giant panda, unlike these animals, does not conjure up metaphorical meanings. Even the references scholars have found that very likely do point to the giant panda in ancient texts do not imply that the panda was of central symbolic importance. Moreover, as Donald Harper, professor of ancient Chinese civilization, notes, even these references were "forgotten in modern times."[4]

Once virtually unknown, even in its native country, the giant panda grew in recognition as a scientific curiosity that challenged western classification systems and basic assumptions about evolutionary biology. Its bone structure, for instance, was distinct from that of a typical bear, yet it did not easily fit into any other group. After the People's Republic of China was founded, the giant panda, as a scientific puzzle and endemic species, quickly became a tool for teaching the Chinese people to embrace modern science and scientific thought. The panda's present-day prominence makes it difficult to believe it was skipped over in most of China's rich literary, cultural, and material history. Yet this elusive animal was skilled at maintaining an unobtrusive existence for many millennia. Since it has emerged into the limelight, its unique combination of scientifically odd and physically attractive features has inspired broad interest. Panda fossil studies have successfully linked the territory of the PRC to its geological past. The panda's lack of recognized cultural meaning made it unproblematic to the modern communist nation, which began to adopt it as a symbol during the late 1950s and early 1960s. The giant panda also rose to an influential position in diplomacy, wildlife conservation, and scientific funding.

David's Discovery

Père Abbé Armand David introduced the giant panda to the Western world in 1869, during his tenure in China as a French-Catholic missionary and priest. The story of Père David's "discovery" of the giant panda has been widely celebrated in popular and scientific writings alike. The giant panda was one of more than 100 creatures that Père David introduced to European science during his combined mission in China as a Lazarist missionary of the St. Vincent de Paul order of the Catholic Church and as a collector of flora and fauna for the Museum of Natural History in Paris.[5] When he obtained a giant panda (not yet named) in 1869, Père David did not consider it the most intriguing of the exotic specimens he found in China.

In his writing, Père David expressed more interest in *Elaphurus davidianus,* a deer-like creature known for its unusual appearance now commonly called "Père David's deer." In Chinese, the animal was called *sibuxiang* (四不象, "four dissimilarities") and was described as having a body shaped like a horse, turned-up hooves, a donkey tail, and ill-suited antlers. Another description of it was that of an animal with the "tail of a donkey, antlers of a deer, neck of a camel, and hooves of an ox."[6] This animal was clearly a taxonomic puzzle. When Père David went to China during the mid-nineteenth century, the *sibuxiang* was already considered to be extinct in the wild and known only to exist within the protected walls of the emperor's imperial hunting grounds. Père David, fascinated by the stories he had heard of this creature, went to great lengths to steal a peek at this odd animal. Once he saw it, he could not overcome his desire to send some specimens back to Paris. After great effort and expense, he managed to obtain a few of these creatures and have them transported live from China to France.[7]

The giant panda, on the other hand, was not as conspicuously curious. Upon seeing the pelt of a giant panda (locally called *bai xiong,* which literally translates to "white bear") in the home of a landowner in Sichuan Province, Père David referred to it as the "famous white and black bear."[8] He sent out hunters to capture one as a specimen for him. The creature's peculiarities with regard to bone structure, diet, and mating practices were not apparent at the time. Père David casually examined the *bai xiong* specimen that his hired hunters had shot and brought to him, classified the giant panda as a bear, and gave it a species name for its most distinctive feature, its black-and-white coloring: *Ursus melanoleucus* (*Ursus* being Latin for "bear" and *melanoleucus,* meaning "black and white"). David wrote a report on his "discovery" and sent it off with the *Ursus melanoleucus* specimen to the Museum of Natural History in Paris in 1870.[9]

The portrayal of Père David's introduction of the species as the "discovery" of the giant panda was an overt romanticization of actual events. David's discovery, like many other discoveries of that era of Western exploration, was clearly a case of transcultural communication: local people introduced a local animal species to a European, who in turn introduced it to European scientific circles. A paucity of historical records relating to the giant panda makes it extremely difficult to determine the extent to which this animal was known within China before 1869. At the very least, it was well known by local people as something that might be of interest to Père David when he questioned them about the region's flora and fauna. That this animal would be valuable to the foreign missionary was quickly perceived, and the "Christian hunters" working with Père David sold it to him "very dearly."[10]

Eluding the Ancients and Confounding the Moderns

Once the "black-and-white bear" became more renowned in the twentieth century, Chinese scholars and panda biologists began to comb ancient texts and various art forms for early observations of the giant panda.[11] The challenge in locating panda citations is to find descriptions that are actually somewhat consistent with characteristics of the giant panda. Ancient texts offer a number of possible references to the giant panda, but they are limited and often vague or highly inconsistent with current knowledge of the giant panda. Among the texts that conceivably included observations of the giant panda is the *Shanhai Jing* 山海经 (The Classics of Seas and Mountains) from the Eastern Zhou dynasty (770–221 B.C.E.), which mentions a black-and-white bear that eats metal. It is doubtful that any animal actually ate metal, but most animals probably would lick a metal pot that had contained food. This detail could also simply be a means of adding to the mythical quality of animals with which people had little contact.

Such ancient texts as the *Shi Jing* 诗经 (Book of Songs) from the Western Zhou dynasty (1050–770 B.C.E.) make reference to several animals, including a mythical panther-like, or possibly bear-like, creature called a *pixiu* 貔貅.[12] Although panda researcher Hu Jinchu and biologist and essayist Zhou Jianren were convinced that this term was an ancient name for the giant panda, too many aspects of the descriptions are counter to giant panda traits for *pixiu* to be a clear reference. According to the classic reference *Er Ya* 尔雅 from the Qin dynasty (221–207 B.C.E.), the term *pi* 貔 more consistently referred to a large cat and was defined as "white leopard" at the time.[13] In his careful review of potential ancient references to the giant panda, Donald Harper's final assessment of the

FIGURE 1.1 This is an image of a Malayan tapir. It is believed that the Chinese term for this animal, mo 貘, referred to the giant panda in ancient texts. Some believe that the animals were simply mistaken for each other; notice the similarity in color and size. The application of this term to both animals, however, might not be based on visual similarities at all. Courtesy of Smithsonian Institution Archives. Image # SIA Acc. 14-167 [NZP-0704].

term *pi* 貔 was that it could not be a reference to the giant panda in part because citations containing this term referred to regions of China in which there is no evidence pandas ever lived.[14]

Hu Jinchu joined other scholars in asserting that the term *mo* 貘 was another ancient term for the giant panda. Hu calls it the "southern term" for the giant panda.[15] The modern translation for the term *mo* is "tapir," referring specifically to the Asian species *Tapirus indicus*. The home range of the giant panda and the tapir apparently overlapped in prehistoric times. Their coloring is strikingly similar and their size is comparable—leading some scholars to believe that the two animals may simply have been mistaken for one another in casual observations during the ancient period. See Figures 1.1 and 1.2 to compare these animals. This would explain why the *mo* 貘 was sometimes linked to very panda-like descriptions as in the previously noted *Er Ya* 尔雅, which defines the *mo* 貘 as an animal that "resembles the bear, small head, short legs, mixed black and white."[16] In other texts like Li Shizhen's *Bencao gangmu* 本草纲目 (Materia

FIGURE 1.2 A giant panda is juxtaposed with the Malayan tapir to compare their appearance. Photograph by J. Matthew Diffley, Wolong Panda Center, Wolong National Nature Reserve, Wenchuan County, Sichuan Province, PRC, February 2002.

Medica), (1596) additional features such as an "elephant trunk" seem to better fit the tapir.[17] The term *mo* 貘 constitutes the core of Harper's detailed investigation into the presence of tapirs and pandas in ancient texts. He contends that *mo* 貘 did refer to the giant panda in ancient texts and that the association between this term and the tapir was actually the result of a mistake made by the nineteenth-century French scholar, Jean-Pierre Abel-Rémusat.[18] Harper explains that the Tang-era poet, Bo Juyi, described a fantastical creature that he called *mo* 貘, which had an "elephant trunk, rhinoceros eyes, cow tail, and tiger paws." This animal was painted on a screen he kept by his bed. Harper notes that this description was subsequently reproduced in a variety of Chinese cultural materials, from woodcuts to dictionaries. Abel-Rémusat examined these in close proximity to the exciting discovery of the Malayan tapir and consequently assumed that the tapir also existed in China and that these texts and illustrations must be references to it.[19] As a result, Abel-Rémusat and subsequent scholars began to define *mo* 貘 as tapir. Although fossils of *Tapirus* that date back to the Pleistocene Era have been found in the same regions of

China as fossils of giant pandas and their ancestors, according to Harper, there is no paleontological evidence of the existence of the tapir in China since well before the Shang dynasty (approximately 1600 B.C.E.–1046 B.C.E.).[20] Harper concludes that since there is no record of the existence of tapirs in China since the use of script, references to the *mo* 貘 prior to the Bo Juyi's poem refer to the panda, any similar subsequent descriptions refer to a mythical creature, and explicit definitions of the *mo* 貘 as a tapir are actually a mistake.[21] If it is true that tapirs disappeared from China prior to the use of script, their similarity in size and coloring is moot with regard to the historical record but has probably contributed to more recent confusion and the assumption by many modern readers that these two animals were conflated in ancient texts.

More recent texts are devoid of even remote allusions to the giant panda. The "black-and-white bear" is conspicuously absent from the records of China's fauna that were detailed in the rich compendia of medical texts dating back to the Song dynasty (960–1279). Harper convincingly argues that Abel-Rémusat's mistake led to the erasure of knowledge of the giant panda's presence in ancient texts, certainly among western scholars.[22] For Chinese scholars, it seems that this erasure had already occurred in the medieval texts and late imperial texts that Abel-Rémusat examined. As Harper notes, a description of the poet Bo Juyi's fantastical beast "had been recorded in *materia medica* compiled by Su Song 蘇頌 (1020–1101)."[23] He also notes that this description was then repeated in the even more famous and more recent collected record of medicinals, the *Bencao Ganmu* 本草纲目 (Materia Medica), compiled by Ming dynasty (1368–1644) author Li Shizhen in 1596. The *Bencao Ganmu* includes information on over 400 medicinal materials derived from animals; not one of these was derived from the giant panda.[24] Even in more comprehensive Qing dynasty (1644–1911) encyclopedias such as the *Gujin Tushu Jicheng* 古今图书集成 (Compilation of Books and Illustrations, Past and Present) from 1725, the giant panda is not found among the extensive listings.[25] Although the giant panda seems to have made its way into some ancient texts, its presence was not lasting. In the extensive efforts by Chinese scholars to amass and catalogue knowledge about the natural world that surrounded them from the Song dynasty on, the giant panda escaped notice, possibly in part due to the fantastical creature that took its name.

An additional explanation for the giant panda's lack of prominence in historical texts is the likely lack of interaction between these extremely reclusive animals and humans. Accounts of twentieth-century expeditions focused specifically on the giant panda affirm that the creature was very elusive. Ernest Henry Wilson, a naturalist who traveled in Sichuan Province and other parts of western China in 1913 collecting and cataloging flora and fauna, called the giant panda the "sportsman's prize above all others worth working hard for in Western China."[26] During the several months that Wilson spent in panda habitat, he did not so much as catch a glimpse of a giant panda and speculated that "the savage nature of the country it frequents renders the possibility of capture remote."[27] When the Chicago Field Museum of Natural History commissioned Theodore Roosevelt's sons, Ted and Kermit, to obtain a giant panda specimen in 1928, they were so skeptical about being able to successfully hunt a giant panda that they generally kept the object of their quest a secret.[28]

> The Golden Fleece of our trip was the giant panda . . . From time to time such men as General Pereira, Ernest Wilson, Zappey, and McNeill had hunted it, but without success. We had slight hope of getting it. So slight in fact that we did not let even our close friends know what our real objective was.[29]

The Roosevelt brothers spent over a month in 1928 trekking around in the mountains of Sichuan before spotting a giant panda.[30] This hunt brought them glory as the first foreigners to shoot a giant panda. At that time, the giant panda was seen as exotic, but not benign or endearing.

Softer associations with the panda only developed when Ruth Harkness, the widow of animal trapper William Harkness, responded to her husband's sudden death in China by going to China to realize his dream of trapping a live giant panda. Ruth Harkness taught the world to adore the giant panda by returning to the United States in 1936 from a two-week-long expedition with a harmless-looking live panda cub cuddled in her arms.[31]

The efficiency with which Ruth Harkness completed her expedition proved to be rare. Rival trapper Floyd Tangier-Smith was extremely frustrated when Harkness succeeded before him in spite of his years of effort. When the Chinese government gave biologist Ma De an assignment to find a panda for the London Zoo in 1945, the huge entourage that was put to the task searched for five weeks before even seeing a panda. They then had to chase it for another three weeks before capturing it.[32] Even George B. Schaller, who pioneered the first major collaborative giant panda research project with the World Wildlife Fund (WWF) and the Chinese government in 1979, spent over two months tracking pandas before seeing his first live panda in the wild.[33] In light of the fact that such focused efforts resulted in so little contact between humans and giant pandas, it is little wonder that earlier chance meetings with giant pandas would result in such sparse and vague descriptions in the historical records, even if pandas were more abundant in earlier times.

Even after the giant panda attracted interest outside China, local attitudes toward the giant panda indicate that people who lived in its habitat had little use for it. Its meat certainly was not prized. Ted and Kermit Roosevelt did not complain as they feasted on their exotic game, but they observed that their local companions were not interested: "not an omnivorous Lolo of the lot would touch a morsel of the flesh."[34] Even a poor peasant who accidentally killed a giant panda in his trap in 1983 within the Wolong Nature Reserve confessed during his trial for illegal trapping that he had tried but discarded the panda meat. "I carried it [the meat] home and my wife cooked it with some turnips. We ate some. It did not taste good, so we fed it to the pigs. And I took some to my sister."[35]

The naturalist Ernest Wilson also noted that local people did not hunt the panda. Its pelt was occasionally mentioned in travelers' accounts as being used as a rug or to sleep on. If the pelt referred to in these ancient texts was actually that of the giant panda, lying on it was thought to be good for staving off dampness and controlling (or stopping) menstruation.[36] Such benefits, however, were apparently insufficient to warrant the energy and time that it took to track down this elusive animal. According to members of the minority Baima group, who live in Sichuan near the present-day giant panda reserve of Wanglang, it was an old belief among their people that killing giant pandas could bring bad luck because they were gentle creatures that simply ate bamboo and did not threaten people.[37] Only after the giant panda became an endangered species did its pelt fetch a price substantial enough to motivate hunters. During the 1980s, panda pelts could bring in tens of thousands of dollars.[38] In recent years, giant panda pelts are said to sell for US $100,000 to 200,000 in Japan. This potential fortune continued to motivate hunters despite the risk of the death penalty, which was inflicted on panda poachers in 1987.[39] Even as recently as 2008, a woman was arrested for attempting to sell a panda pelt for a purported US $38,000.[40]

Inaccurate descriptions of the giant panda's behavioral habits offer further evidence that local people did not have frequent contact with and intimate knowledge of this animal. Based on second-hand knowledge, Ernest

Wilson described the giant panda as a hibernating animal. He wrote, "According to the natives, the Peh Hsiung (giant panda) hibernates for six to seven months in the hollow trees, dry rocky hollows and caves."[41] Recent behavioral research contradicts this information. Because giant pandas must invest so much time eating bamboo in order to consume enough calories for their sustenance, they do not have enough time or fat storage to hibernate.[42] The contrary "observation" by these local people demonstrates that at most they had only casual contact with these animals. Another explanation for this discrepancy between early accounts and present-day observations is that the local people who related this information to Wilson actually described pregnant pandas, which remain in a secluded spot for a few months to birth their offspring. Either the observers themselves or Ernest Wilson may have misinterpreted this behavior as hibernation. This more recent evidence of inaccuracies about the giant panda reinforces the notion that the vagueness of the ancient references to panda-like creatures are due to a lack of intimate interaction with these animals. Even Père David's basic observations of the giant panda specimen would not have taken human knowledge of this animal much further had his colleague, Alphonse Milne-Edwards at the Museum of Natural History in Paris, not looked more deeply into what it had to offer the scientific community.

Solving the Taxonomy Debate

Père David knew this handsome "black-and-white bear" would be of interest to science, but he did not realize that his "discovery" would lead to the finding that the animal might not be a bear. This subsequent revelation occurred in Paris when Alphonse Milne-Edwards performed dental and osteological exams on the *Ursus melanoleucus* specimen that Père David sent to him and determined that it was strikingly similar to a member of a different animal family, the panda (now known as the red panda or lesser panda). It soon became clear that David's "*Ursus melanoleucus*" did not fit easily into Linnaean taxonomic categories. According to Milne-Edwards, the "black-and-white bear" was not a bear at all, but rather part of *Procyonidæ* (the raccoon family).[43] Naturalist Thomas Hardwicke had "discovered" the (red) panda in the Himalayas in 1821, and zoologist Frédéric Cuvier then classified the small, rust-colored, ring-tailed animal as *Ailurus fulgens* (bright, cat-like animal) in 1825, see Figure 1.3. Cuvier in turn gave the *Ailurus fulgens* the common name "panda," which was derived from its Nepali name.[44]

FIGURE 1.3 It is possible to see why this red panda might be grouped with raccoons; its likeness to the giant panda is less obvious through casual observation. Photograph by J. Matthew Diffley, Chengdu Research Base for Giant Panda Breeding, Chengdu, Sichuan Province, PRC, February 2002.

In an effort to represent the affiliation that he saw between the "black-and-white bear" and the Himalayan panda, Milne-Edwards created the new genus *Ailuropoda* (cat-footed, or with a foot like that of the *Ailurus*), which he later altered to *Ailuropus* (like *Ailurus*, or like the panda). He removed the giant panda from the *Ursidæ* (the bear family) and placed it in the *Procyonidæ*.[45] Milne-Edwards retained David's species name, *melanoleuca,* which described its coat coloration. Previously, the red panda simply had been called "panda," but after the large black-and-white animal in the same family was classified as a panda, a distinction had to be made. The much larger "black-and-white bear" became the "great panda" and then the "giant panda," and the smaller animal was called the "lesser panda."[46]

Milne-Edwards' reclassification of this "black-and-white bear" not only determined its present-day Latin and common references, but also the modern Chinese name for the giant panda, *da xiongmao* 大熊猫, literally translated as "great bear-cat." Prior to this foreign taxonomical intervention, people who dwelt in the home range of the giant panda most commonly referred to it as *bai xiong* 白熊 (white bear).[47] In some regions it was also called *hua xiong* 花熊 (spotted or patterned bear).[48] People who presently dwell in the vicinity of the panda's home range continue to use these local terms, but they also recognize the standard dictionary term, *da xiongmao*. Like the English terms, the Chinese terms for these two animals distinguish the two animals by size. The Chinese terms are a clever fusion of the Latin and common names that emerged from Milne-Edwards's conclusions. The inclusion of *xiong* 熊 (bear) in both names reflects the Chinese names of both animal families, *Ursidæ* and *Procyonidæ*, central to the taxonomic debate; the Chinese term for *Ursidæ* is *xiongke* 熊科, or bear family, and for *Procyonidæ* is *huanxiongke* 浣熊科, or raccoon family. The Chinese terms also group the two animals together as being the same basic type. This nomenclature did not, however, inspire or reflect a consensus on the classification of these animals within China.

When Milne-Edwards renamed *Ursus melanoleucus* as *Ailurus melanoleuca*, he triggered an extended taxonomical debate. After various specialists offered a range of opinions, he altered his stance and by 1874 promoted placing the giant panda in a taxonomic position between the bears and the lesser panda, but did not advocate the creation of a new family.[49] The debate continues today in the scientific literature. Despite the application of recent molecular scientific advancements and the benefit of new developments in DNA analysis, the question of "panda" identity remains unresolved.[50] In creating a classification puzzle the year after its "discovery," the giant panda became a significant scientific oddity. An animal originally striking only because of its unusual coat became controversial for what lay beneath—its skeletal structure. Later studies on the physiology, feeding, mating, and other behaviors of the giant panda only further confused its taxonomy. The giant panda inspired a host of new questions in the scientific community over which features should be most relevant for taxonomical classification.

The giant panda historically challenged taxonomic conventions because its traits fit multiple categories but made it a misfit in any single one. Numerous mammalogists examined and reexamined the skeletal and dental structure of this animal and then frequently reclassified it. Two scientists, E. R. Lankester and Richard Lydekker, emphasized the affinity between the "black-and-white bear" and the red panda and were the first to call it *Ailuropoda melanoleucus* or the "great panda." They posited a greater affinity between the two pandas than between either of the pandas and other animals and placed them in the family of *Procyonidæ*, with individual subfamilies for each panda species.[51] Various schools of thought formed around the debate, and both types of pandas were simultaneously classified in multiple and conflicting categories.

For Milne-Edwards, the giant panda skull provided the original evidence that the giant panda was not a typical bear. Fossil skulls added more evidence that the giant panda was even odder than originally thought. A group of scientists found Late Pliocene mammalian fossils in Sichuan in 1921, and the skull fossils that paleontologists W. D. Matthew and Walter Granger examined seemed strikingly similar to those of a giant panda. They concluded that they had the giant panda's ancestor in their hands. They also concluded that the giant panda was indeed a bear, but not a true bear.

> The *Æluropus fovealis,* new species. The teeth resemble those of *Æluropus melanoleucus* [the giant panda]. . . . The affinities of *Æluropus* appear to be with *Hyænarctos* [an extinct separate branch of the bear evolutionary stem]. . . . Its systematic position appears to be clearly in the family *Ursidæ,* although of a distinct subfamily from true bears. The occurrence of *Æluropus* almost completely modernized in the Pliocene, if these deposits are in fact Pliocene contemporary, or nearly so . . ..[52]

Matthew and Granger compared these fossils to Lankester's illustrations of a giant panda skull and teeth in order to draw their conclusions. Lankester had maintained the theory that the giant panda belonged to the *Procyonidæ,* while Matthew and Granger used Lankester's illustrations to determine the giant panda was a bear. Their Pliocene fossil appeared to be a bear, albeit an unusual one, and the giant panda's similarity to this animal reclaimed it from the raccoon family and returned it to that of the bears. As more evidence surfaced and more studies were carried out, the taxonomy debate thickened.

D. Dwight Davis, curator of vertebrate anatomy at the Chicago Natural History Museum, accused Western participants in the panda debate of basing their conclusions on political or geographic allegiance rather than on scientific rigor. He blamed the lack of consensus among experts between 1870 and 1964 on two factors:

> The data employed [were] not sufficient to form a basis for an objective conclusion . . . opinion as to the affinities of Ailuropoda is divided almost perfectly along geographic lines [of the scientists],

which shows that authoritarianism rather than objective analysis has really been the determining factor in deciding the question.[53]

Davis clarified this "geographic" distinction by explaining that a consensus could be found among British and American scientists, who ultimately placed the pandas in *Procyonidæ* (the raccoon family). In contrast, the continental European school of thought generally placed the giant panda in *Ursidæ* (the bear family). Davis challenged the objectivity and universality by which scientists defined their work, highlighting the turmoil this odd animal generated in the scientific community.

Davis argued that only through more extended analysis of more varied data could the giant panda's classification be determined. He personally provided this with his own morphological study of the giant panda in 1964. Stephen Jay Gould, evolutionary biologist and historian of science, commended Davis's study saying, "[It is] probably the greatest work of modern evolutionary comparative anatomy, and it contains more than anyone would ever want to know about pandas."[54] After completing his exhaustive study, Davis concluded that "the giant panda is a highly specialized bear."[55] In doing so, he broke free from the geographic-linguistic trend he had observed and abandoned the Anglo-Saxon championing of the raccoon classification for the continental classification of bear. Subsequently, scientific opinion began to fall in line with Davis's conclusion, but the issue was by no means put to rest.[56] George B. Schaller, after reviewing a century of investigation into the taxonomy of the giant panda, surveying studies that used highly advanced technology such as DNA analysis, and conducting personal observation of giant pandas in the wild, arrived at virtually the same conclusion in 1993 as Milne-Edwards had in 1874: "the panda is a panda."[57] A 2007 phylogenetic analysis of the giant panda's mitochondrial genome by a group of Chinese scientists concluded that the giant panda is more closely related to bears than to the red panda, but it is not conclusive on the question of whether or not the giant panda should be included with *Urisids* or be placed in a sister group.[58] Olaf R. P. Bininda-Emonds, head of the Bioinformatics Research Group at the Technical University of Munich, completed a super-tree analysis that incorporated 105 studies that took stances on the taxonomy of the giant panda and the red panda and included an analysis of diachronic trends in studies from various scientific disciplines. He noted that the scientific discipline had strongly influenced the conclusion; for instance, behavioral studies highlighted commonalities between the giant panda and the lesser panda, while molecular studies overwhelmingly favored grouping giant pandas with bears.[59] Considering that new techniques, new types of data, and new systems of classification have kept this debate alive, Milne-Edwards should really be considered the one who, very literally, discovered the giant panda by looking beneath the surface of its unusual coat and introducing a puzzle in its skeletal structure that has intrigued scientists ever since.

Fossil Findings

Scientific research on giant panda fossils reveals that the animal's range was previously much wider than it is today. In 1915, paleontologist A. Smith-Woodward reported on a skull found in Burma that he determined belonged to a relative of the giant panda. This was the first clear indication that the former range of the giant panda and its ancestors extended far beyond the animal's present confines in the mountains of Sichuan, Shaanxi, and Gansu Provinces.[60] Pleistocene pandas lived throughout southern China, extending as far south as present-day Vietnam and as far northeast as Beijing.[61] Reports in both Chinese and Western languages, ranging from elementary introductions to highly technical papers on fossils, mention the wide expanse of the giant panda's early territory. Refer to Figure 1.4 to compare the giant panda's current range with its prehistoric range based on panda fossils. Its broad prehistoric range helped validate the giant panda as a national treasure rather than simply a regional animal of potential national interest. Most Chinese sources published in the 1980s or earlier fail to mention that giant panda ancestral fossils were also found in present-day Vietnam and Burma as early as 1915, even though their references often contained studies that described these "non-Chinese" giant panda fossils.[62] This puzzling omission, however, reflects a nationalistic concern, possibly on the part of Chinese publishers, that China be allowed to claim the sole origin of the giant panda. A popular Beijing magazine passionately rebutted Western scientists' claims that a panda tooth fossil older than the fossils found in China was purportedly found in Eastern Europe and thus proved the animal had European origins.

In an effort to reclaim Chinese primacy, the magazine cited the work of Chinese paleontologist Qiu Zhanxiang, who had discovered a much older fossil within present-day China that could be linked to the giant panda.[63] Qiu and his colleagues discovered an amazing trove of fossils near Lufeng in Yunnan Province, where they found fossil teeth that they determined

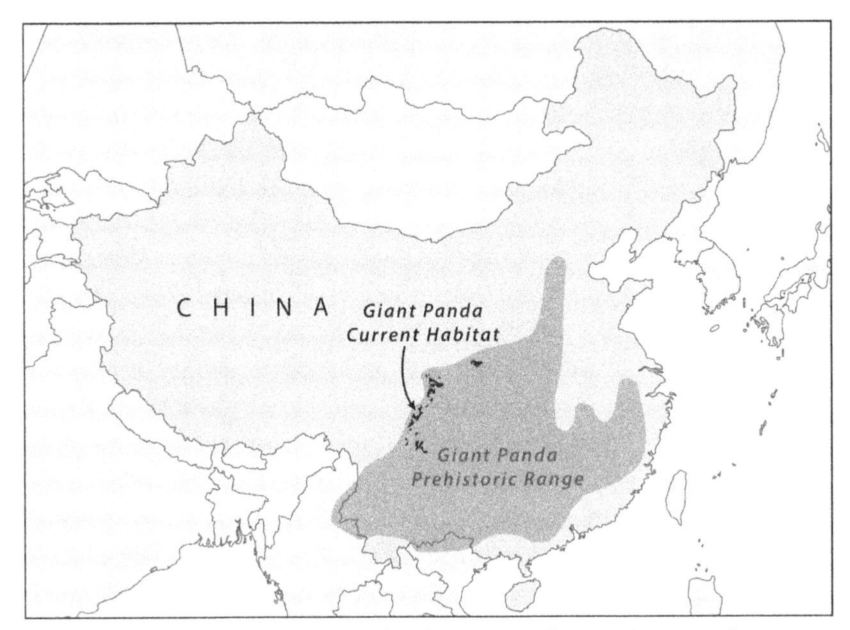

FIGURE 1.4 This map illustrates the dramatic difference between the current range of the giant panda and its prehistoric range based on fossil records. Map created by David Medieros, 2015.

belonged to a direct ancestor of the giant panda, which they named *Ailurarctos lufengensis*.[64] This discovery enabled Qiu to link the giant panda to an earlier species dated to approximately 7–8 million years ago (the late Miocene epoch).[65] In 2007 another early panda skull was found in China, but the 2012 discovery in Spain of skull fragments of what might be an ancestor to the giant panda again challenged the giant panda's "China" origins.[66] This need to assert the giant panda's Chinese origin may seem somewhat ludicrous in light of the fact that China did not exist during the Miocene and Pliocene epochs when these now fossilized animals were roaming Earth. The desire of present-day citizens to forge a sense of continuity between the territory of the present nation and the same region millions of years ago through the giant panda embodies one of the quintessential tensions of the modern nation—that legitimacy requires both a claim on the past and a break from it. While the break from the past is usually dominated by an effort to make political distinctions from former governing systems, efforts to simultaneously claim the past manifest themselves in many forms, including maintaining historic traditions, preserving ancient ruins, and showcasing geological features, flora, and fauna that predate human existence.

The effort to link China to the geologic past also is reiterated in discussions surrounding the unusual lack of evolutionary change that the giant panda has undergone during its long existence with the repeated characterization of it as a living fossil, or *huo huashi* 活化石. This clever paradoxical phrase, implying that a living animal is highly primitive and thus a window into the past, conjures up a powerful image. As an actual phenomenon, however, so-called living fossils are not uncommon. Giant pandas could command greater interest than other "living fossils" because they are qualitatively more distinct and quantitatively rarer than most other extant primitive animals in the modern world. In terms of evolution, many extant crustaceans, fish, and insects predate dinosaurs. For land mammals this status is more noteworthy, yet the fossils of such present-day mammals as the opossum, Laotian rock rat, and okapi demonstrate that they were also, as Matthew and Granger so cogently phrased it, "completely modernized in the Pliocene."[67] Each time the panda is presented as a "living fossil," however, it is described in tones of respect and awe. In China, discussion of the panda as a "living fossil" is frequently accompanied by the phrase "national treasure," *guo bao* 国宝, or one of the terms used to refer to endangered species, *zhenxi dongwu*, 珍稀动物. Naming the giant panda a living fossil, a notion repeatedly refuted by many scientists, adds a primordial quality to the panda's rarity, making it all the more precious.

The survival of the giant panda species through the Ice Age is portrayed in popular literature as something akin to a miracle, adding to the mystery of the panda's existence. The giant panda's minimal evolutionary change inspired Stephen Jay Gould to write several essays that characterize the animal as an oddity, even a fluke. Gould argued that such unusual characteristics as the giant panda's bamboo diet were not clever adaptations, but impediments to its survival, which seem to have distinguished the panda as an even more amazing animal. He writes: "The primary theme of panda life must be read as a shift of function poorly accommodated by a minimally altered digestive apparatus."[68] For Gould, the close structural affinity between the present-day giant panda and its fossil ascendants is a prime example of the serendipity of evolution. In his view, even the panda's famous thumb (commonly seen as an ingenious adaptation) that allows it to manipulate bamboo stalks with surprising dexterity did not develop from the digit that would have been most ideal for the job.[69] The giant panda is the perfect example of haphazard, even undesirable, transformations that challenge both the existence of divine direction and the teleological view of evolution that all present features arose for their express functions. Although Gould's agenda in examining giant panda evolution is detached from the goals of Chinese scientists, his essays offer an external view that this evolution, or lack thereof, is noteworthy and reinforces the wonder of its modern existence.

Much of the attention paid to the primitivity of the giant panda resulted in broad glossing over of the interesting species diversification discovered in panda fossils. Recent scientific studies reveal that there were a number of different early panda species. After analyzing what they determined to be the earliest panda skull, Changzhu Jin, Russell L. Ciochon, and their colleagues note that this Pliocene epoch panda fossil is the skull of the *Ailuropoda microta*, a species notably smaller than the modern *A. melanoleuca* and an ancestor to the present living panda. Furthermore, at least two other panda species existed during the Pleistocene epoch (1.8 million to 11,000 years ago), namely the *Ailuropoda wulingshanensis,* dating back to between 1.65 and 1.95 million years ago, and the *Ailuropoda baconi*, which can be dated to about 750,000 years ago.[70] Scholars have not been able to determine when the present living panda *Ailuropoda melanoleuca* came into being. Gaps in the fossil record allow for a number of possibilities, including the coexistence of the present-day *A. melanoleuca* and other giant panda species during the Pleistocene epoch.[71] Specifically, Jin and Ciochon note that "a preference for a diet of bamboo has characterized the lineage since the late Pliocene" and furthermore that "the new specimen established that the cranial anatomy of the giant panda (and probably the postcranial skeleton) remained essentially uniform, except for size and minor dental alterations, for more than 2 Myr [million years] during the late Pliocene and Pleistocene, a period of pronounced global climatic instability."[72] Even as this recent discovery refutes the broadly implied notion that the present-day giant panda existed as far back as the Pliocene, Jin and Ciochon's research concurs with the argument that the degree of evolutionary change that occurred between this earlier panda species and the present-day one was minimal.

Both the long duration of the panda's existence and the wide expanse of the panda's range within China make its position as a national symbol more thoroughly grounded and powerful. The scarcity of historic textual references to the panda actually makes it an even more ideal symbol for the modern nation. Because the giant panda did not embody blatant or broadly recognized traditional meanings, the PRC government did not have to overcome what it might consider unsavory links to the culture of the imperial era in order to enjoy the benefit of the panda's geological ties to the geographical region of China. Such positive and nation-reinforcing connotations, combined with the panda's physical appeal, contributed to the production of this animal as a symbol that could attract both official endorsement and widespread popularity at home and abroad. In portraying the living panda as having natural ties to the land that reach back millions of years, scientists, advertisers, and politicians alike have transformed the panda into a bridge between the present and a very distant past. The panda thus proves to be a potent symbol for seemingly contradictory elements of the modern nation: it represents a unique link to the past and a vehicle for scientific engagement with other nations in the present day.

The Panda Promotes Communist Science

After the founding of the PRC, the new government saw science as a means of both liberating and strengthening China as a nation. From the first month of its rule, it actively organized such scientific institutions as the Chinese Academy of Sciences (CAS), which governed six research institutes plus the Institute of History and Languages.[73] Such early and aggressive political assertion of control over science demonstrates the government's view of these organizations as both potentially powerful and potentially threatening. These concerns and their reflection in the close relationships among the state, the military, and the scientific establishment are of course not unique to China, communist, or authoritarian governments.[74] This close coordination between scientific research and the state was seen by the Chinese government as a new and important element of the modern state. As historian of science Shuping Yao noted, "From the outset the CAS stressed that, unlike the situation in the scientific research establishments of Old China, scientific research must now promote national economic development."[75] The association between the central government and scientific research on the giant panda during the early Communist period was likewise neither casual nor coincidental. While panda science offered no conspicuous economic application, the sudden emergence of scientific research on the giant panda in China during the 1950s illustrates the Communist Party's recognition of its relevancy to the state.

Prominent Chinese biologist Zhou Jianren drew strong links between science, pandas, and the nation in his 1956 article, "Guanyu Xiongmao"

(On Pandas), in the *Renmin ribao* (人民日报 or *People's Daily*), the mouthpiece of the Chinese Communist Party.[76] Zhou emphasized the particular utility of the panda as a tool for training the Chinese people to embrace a more scientific way of thinking.[77] He stated, "These animals [the pandas] presently have a special function. Doing research on them can increase our respect for objectivity. The study of most natural science has this use, but it can be said to be especially true with the study of the panda."[78] For Zhou, the panda served as the perfect object to promote the benefits of scientific research because the only way to understand it was through methodical study and field research. In his view, simple first-hand observation was not powerful enough to unveil the subtle curiosities of this animal. As a scientific puzzle, the panda was characterized by behavior that countered the logic of its physical structure. The panda would foil any attempts to understand it by the "old" method of knowledge acquisition, meaning, according to Zhou, "casual" observation.[79] Its paradoxical classification as a carnivore and its behavioral habit of eating only bamboo required that it be researched with rigorous scientific methodology: "Because the panda is categorized as a carnivore yet does not eat meat, if one simply depended on common judgment, it would be difficult to avoid making a mistake."[80] Zhou thus used the scientific study of the panda to promote what he called "scientific thinking," in contrast to the behavior of "ancient people" who were not "looking at details and were making baseless assumptions."[81] Zhou stressed the role of modern scientific thinking in state efforts to eradicate superstition and lax behavior during the 1950s. In this way, the giant panda became a weapon for fighting against "feudalistic thought" and "backward" behavior.

Positing a stark contrast between "old" and "modern" ways of thinking, Zhou advocated the style of thought he characterized as scientific and superior. His underlying agenda was to build a stronger nation. According to his logic, if all of China's people could master this modern, scientific perspective, China would naturally become a stronger country. Thus, just seven years after the inception of the People's Republic of China, the giant panda was already symbolizing the science-centered ethos of the time.

Zhou's 1956 newspaper article reflects the broader project of "national science" during the Mao Era. The same year that Zhou's article was written, Mao Zedong explicitly promoted expansion in scientific learning and incorporated scientific work into the First Five-Year Plan.[82] Promoters of scientific learning created a movement in order to involve more people in the study of science and promote the application of scientific methods to other sectors such as agriculture and industry.[83] The privileging of science went beyond teaching people to think logically and systematically. It dictated the allocation of funds and promoted new projects that showed promise for scientific advancement. That same year, the Third Plenum of the First National People's Congress approved a proposal by the Chinese Academy of Sciences to create a nature reserve for ecological research. While at the time this project was not directed toward giant pandas, it was inspired by the same scientific fervor that had characterized Zhou Jianren's article, "On Pandas."

In addition to serving as a tool to train the Chinese people to be modern thinkers, the panda also enabled scientists in socialist China to take the lead in the realm of panda science. A 1963 Chinese newspaper article about the world's first successful birth of a giant panda in captivity at the Beijing Zoo claimed that the panda's entrance into the modern world was related to science, stating, "In 1869 it [the giant panda] was discovered by a scientist."[84] This curious acknowledgement of Père David's role, while omitting the fact that he was a French priest, was striking because it conspicuously avoided crediting this relationship to a foreigner and emphasized science as the foundation of the giant panda's significance. As such this article was another reflection of the value being placed on science by society and the government in China at the time.

The successful work of Chinese scientists in panda reproduction also gave China a competitive edge in the future of panda science. The author of this 1963 article pointed out that the birth of this baby giant panda "is extremely valuable to scientific and zoological research." Attesting to the difficulty of this achievement, the article states, "since Liberation, the Beijing Zoo has taken care of more than ten giant pandas for a long duration of time, but until now none have reproduced."[85] Reinforcing the superiority of PRC scientists over their Western counterparts, the article ended by boasting: "In Europe and America there are a number of pandas living in zoos and, although some have tried to breed the pandas, none of them have succeeded yet."[86] The author highlighted panda science as a focal point of Chinese scientific competitiveness in the international arena. In 1963, China was not only able to claim the giant panda as an indigenous species, but was also able to claim superior scientific knowledge of the animal. The article's emphasis on this success belonging specifically to the Communist era tied the claim of superiority in panda-related science to the national project.

From National Allure to National Icon

When the Beijing National Zoo introduced the giant panda to the public with the arrival of the three pandas, Ping Ping, Xing Xing, and Chi Chi, from Sichuan Province in 1955, the Chinese public became familiar with the previously elusive animal and quickly fell in love with it.[87] In 1953, the Chengdu Zoo in Sichuan had been the first in China to display a captive panda, but this animal unfortunately survived for only three weeks.[88] The pandas at the Beijing Zoo proved more resilient and were trumpeted in the national press. News articles about the pandas in Beijing extended the public participation in this phenomenon far beyond the 10,000 visitors that the zoo could expect on holidays.

The first task of the panda display and related newspaper articles was to teach the Chinese people what kind of animal the giant panda was. The inclusion of such basic information as the physical appearance and location of the home range of the giant panda indicated that these general facts were not common knowledge at the time. That the giant panda was unfamiliar to the Chinese public previously was made evident by the appearance of the image of the giant panda as part of a quiz in the youth magazine, *Liangyou* 良友 (Young Companion) in 1940. Using its local name, *bai xiong* 白熊, readers were to guess which region of China the creature was from.[89] The persistent need to describe the giant panda physically over a decade later demonstrates that the animal remained unfamiliar to most Chinese people well into the PRC period. Both the 1956 and the 1963 *Renmin ribao* articles opened by describing the panda as *feipang* 肥胖 (fat), or with *hen yuan de shen* 很圆的身 (very round body), *rubai* 乳白 (milk white) fur, *xiao qiao de yuan de erduo* 小巧的耳朵 (small delicate round ears), and *huaji* 滑稽 (funny, comical), which portrayed the panda as a benign and amusing animal.[90]

Other information considered crucial in the early introduction of the panda included the facts that the giant panda was the zoo's most popular animal, that it was in danger of extinction, and that the panda's native habitat was in western China. The panda's ability to captivate Western and Chinese crowds alike indicated that its appeal was broader than its exotic origins. During periods of Maoist national enthusiasm, the fact that the panda was from China may well have enhanced its popularity in country, but for different reasons. In essence, the panda exhibit at the zoo put China on display for domestic appreciation, like earlier animal displays at China's Imperial Park and Imperial- and Republican-era expositions.[91]

In contrast to Western zoos, Chinese zoo exhibits largely consist of indigenous animals, in part due to financial limitations but also to the wider range of domestic biodiversity. China used the environment to present its own national richness in wildlife to its citizens, more akin to a collection of Emperor's gardens. In the Beijing Zoo, Chinese citizens could view the animals on display as prized examples of the different regions that make up China.

In addition to offering a basic description, the newspaper articles also characterized the giant panda as a likable animal. This sentiment was reinforced by the panda's ability to charm the crowds. In his 1956 article, Zhou Jianren described the zoo visitors as crowding around to see the panda "(because) people really like looking at it."[92] Another newspaper article from 1963 called the pandas by name and described their gentle behavior: "After eating a little snack, the largest one, Ping Ping, lies down."[93] Unlike most zoo animals, all pandas in captivity were named, an uncommon practice at the time for zoo animals in China and internationally. After Ruth Harkness brought the cute panda cub named Su Lin to the United States in 1936, other pandas extracted from China for foreign zoos were all given names, which might be one reason for the appearance of this practice within China.

The Beijing Zoo further anthropomorphized the giant pandas when it advertised the birth of the first captive-born baby panda at the zoo in 1963. Complementing the wholesome and domesticated descriptions of the giant pandas in the 1950s articles, the 1963 birth announcement of Ming Ming, the first captive born giant panda, reported that "the mother and baby are both doing well."[94] In this fashion, the author of the article drew the Chinese public into the "family life" of the new panda in Beijing and effectively made the giant pandas and the Chinese public part of the same national family.

Thirty-six years later on the opposite side of the globe, when the same success in captive breeding was finally achieved in the United States with the birth of Hua Mei at the San Diego Zoo, the giant panda cub inspired the same familial affection. As the product of international cooperative research, Hua Mei made the national family global. In spite of her American birth, however, Hua Mei continued to be a representative of China, where she later returned. In her "native homeland," Hua Mei continued her diplomatic work through reproduction. Chinese officials and staff workers at the Wolong panda breeding center chose Hua Mei's first son, later named Tuan Tuan, to become part of the pair of pandas offered as a controversial gift to Taiwan in 2005 and sent in 2008.[95] Tuan Tuan's suitability for this high-profile position was credited to his pedigree; he was seen as part of "a long line of ambassadors."[96] Even on the international stage decades later, the same basic factors remain prominent in perpetuating the panda's popularity—science, allure, and its exclusive existence within China. The greater its global presence, the more domestic affection it inspires, and the more the giant panda serves as a symbol of China to the world.

2

Nation Building and the Nature of Communist Conservation

ALTHOUGH THE GOVERNMENT of the People's Republic of China values the giant panda as a national possession largely because of its status as a threatened species that is adored worldwide, international concern for the giant panda was not what prodded it to take steps to preserve this animal. In fact, the PRC government recognized the need for protection well before endangered species status was granted in 1990.[1] Efforts to protect the giant panda in the wild first took effect when the PRC was virtually cut off from the rest of the industrialized world, shortly after the Sino-Soviet split of 1960 and well before China's overtures to the international community in the 1970s.

Panda preservation grew out of domestic and scientific concern about sustaining China's natural resources. The scope of nature-protection efforts, policies, and reserve construction during the early decades of the PRC was certainly modest. Only a handful of reserves were created before the late 1970s. However, the very fact that nature protection was legislated at all during these years of early nationhood, a period characterized by mass movements, collectivization, and catastrophe, is remarkable. Inherent in the debates over nature-protection efforts was the basic question of what role nature should play in the construction of a young socialist state.

Although Chinese officials initially took a broad interest in a wide array of sectors and their potential to contribute to the development of Chinese communism, the government soon became more narrowly focused on rapid industrialization and specifically on Mao's campaign of the Great Leap Forward (1958–1959). When pressured by economic downturn following the failure of this campaign and the ensuing famine, Chinese bureaucrats developed a more pragmatic view of nature. The government created a new policy in 1962 that interwove nature and wildlife protection with the establishment of a new hunting industry as part of its effort to respond to the mass-scale crisis. Although the Chinese government recognized the value of studying nature for its potential benefit in the construction of a socialist state during the 1950s, it placed little emphasis on nature protection until dire circumstances during the early 1960s forced it to make nature an integral component of its economic planning. This shift in policy reflected a new appreciation for the material value of nature. The government's simultaneous effort to make nature more productive and to protect precious species that lacked economic benefit, however, demonstrated that officials also grew to value China's wildlife in a way that encompassed nature's symbolic value, as well as its ability to contribute materially to the state. As such, this policy reflected concern for protecting China's aesthetic natural beauty and biological uniqueness.

Protecting Nature for Development

During the first decade of its existence, the People's Republic of China actively looked to the Soviet Union for guidance, funding, and expertise in the broad project of constructing a new state. The Soviet emphasis on the development of heavy industry inspired the Chinese Communist Party to direct scientists to focus their energy on similar endeavors, such as large-scale geological surveys in search of energy resources.[2] In addition to emulating the role of science in the Soviet Union, Chinese scientists in the 1950s established intellectual exchanges in zoology and botany, among other scientific fields.[3] Soviet and Chinese journals exchanged articles, and schools embraced Soviet-style instruction and textbooks.[4] Although this close adherence of scientific research, practice, and teaching to political agendas created serious problems in both the Soviet Union and in China, there were many gifted scientists in both nations who managed to practice good science, gain the ear of the government, and produce environmental policy with both good intentions and beneficial results.[5]

China's close study of Soviet science made it susceptible to the direct importation of some misguided scientific agendas from the Soviet Union, such as Lysenko's theories on heredity, including the idea that new strains of grains could be created by manipulating the genetic response of plants to light and temperature at different phases of their development. Fortunately for China, nature protection in the PRC was spared much of the tumultuous history that Soviet conservation suffered as a result of Lysenkoist science and its rejection of Mendel- and Morgan-based genetics because China's implementation of nature protection coincided with the discrediting of Lysenko-inspired theories in both countries.[6] Nevertheless, some bold notions of control over nature did find their way into Chinese journals. Many nature-protection theories, such as attempting to control a given species through the introduction of an exotic species, had long-term influence over PRC nature-protection policy.

The notion that science was a means of unlocking nature's secrets for the sake of developing society was a prominent theme in scientific writings at the time. A Soviet author featured in the popular magazine *Zhishi jiushi liliang* 知识就是力量 (Knowledge Is Power) wrote an article entitled "Ziran jie li meiyou mimi" 自然界里没有秘密 (Nature Has No Secrets), which argued to a lay audience that science could unravel and dissect earthquakes, volcanoes, and other so-called mysteries of nature from a scientific perspective.[7] By privileging the power of scientific knowledge over such dramatic natural phenomena, the article implied that the state could achieve command over nature. Belief in the ability to unlock nature's secrets reflected the broader scientific movement and promotion of scientific learning in popular society at the time.

The concept of nature protection and the establishment of nature reserves were among the many scientific projects Chinese scientists discussed at length with their Soviet counterparts. The PRC created its first nature reserve, Dinghu Shan, after a group of five members of the Chinese Academy of Sciences (CAS), including Vice President Zhu Kezhen, presented a proposal at the Third Plenum of the First National People's Congress in 1956.[8] This reserve was intended to facilitate research on the interaction between various elements of naturally occurring systems. By endorsing the proposal to establish the Dinghu Shan Nature Reserve in the southern province of Guangdong and placing its management under the jurisdiction of CAS, China's central government embraced the principle that the scientific study of nature was beneficial to the state.[9] Subsequently, efforts were made to set up another nature reserve: Changbai Shan in Jilin Province, located in the temperate region of northeastern China. Changbai Shan's mission was to conduct comparative research on the differences between the temperate climate of Jilin and the tropical climate of Guangdong.[10]

The guiding principles of China's first nature reserves reflected the work of Vasilii Nikitich Makarov, head of the administration of Soviet nature reserves in the 1930s. In one of his last speeches before his death in 1953, Makarov emphasized that the Soviet nature reserves should be laboratories *of* nature, not merely laboratories *in* nature.[11] This distinction promoted the notion that nature should be protected and studied as it was, rather than be subject to experiments that attempted to improve it for the purpose of serving human society. While this approach was criticized in earlier periods in the Soviet Union as bourgeois science, its advocates defended its place in socialist science as the best means of unveiling insights of natural systems and discovering natural organisms and other resources that could benefit society as a whole.

Zhu Kezhen articulated a similar perspective in his description of the plans for China's nature reserves. In an article he wrote in 1959, Zhu noted that plans for nature reserves as sites of scientific research actually began in CAS as early as 1954.[12] Zhu noted that the scientists who formulated these plans hoped that nature reserves would facilitate their research in natural topography, climatology, and hydrology, as well as soil, plant, and animal geography.[13] These plans and the subsequent creation of Dinghu Shan reserve in 1956 clearly reflected the spirit of Makarov's speech. While Zhu Kezhen and his colleagues supported the notion of preserving segments of nature to observe it, they did not consider this to be the only way to study nature, nor did they view the study of nature "as it was" as the only means of using nature for the benefit of the state. The concept that nature could and should be improved upon for the betterment of society was also invoked by the very people who promoted the founding of Dinghu Shan as an observation site. Zhu Kezhen himself did not consider the two views of nature to be mutually exclusive. From his perspective, the two approaches were among the many ways that nature could be used to promote socialist development.

The expansion of nature-reserve research, however, did not progress much after the founding of Dinghu Shan. During the 1950s, concern about the protection of nature was seemingly limited to the scientific community. While the central government recognized that nature protection had a place in the socialist project, it was one of future potential, not immediate need. Many scientists continued to advocate for the expansion of nature-protection efforts; however, the government's inaction indicated that it did not see nature protection as central to its development goals. During the years following 1956, the government was preoccupied with collectivization efforts and the ambitious industrialization project of the Great Leap Forward campaign. Zhu Kezhen and others continued to advocate for nature protection through the 1950s and to articulate the role of nature and nature protection in socialist society along the lines developed by the Soviet Union.

Biologist Jin Jieliu published an article in *Dongwuxue zazhi* (Journal of Zoology) in 1957 that praised Soviet guidance on the integration of nature protection and socialist productivity. He heavily referenced Soviet sources and echoed the connection among science, nature, and socialist production. Jin argued, "Nature protection is highly valuable to the national economy and simultaneously offers limitless cultural, educational, and scientific significance."[14] Jin described science as playing a mediating role between nature and society: "Research on nature in the fields of technology and chemistry enables even more new and useful plants and animals to be discovered."[15] This and similar articles served to underscore the value of such developments as the Dinghu Shan reserve and called for continued government support of the biological sciences and the study of nature.

Later that year, this same journal published an article by Georgii Petrovich Dement'ev, former co-vice president of the All Russian Society for the Protection of Nature (VOOP). Dement'ev traced the development of Soviet nature protection and noted the strong links among nature, science, and the state: "Today, the Soviet Union sees nature protection as a socialist and scientific problem. The organization and management of natural resources must therefore be built on a scientific foundation."[16] Dement'ev's article emphasized that the Soviet Union Academy of Sciences conducted important research through nature parks and also disseminated lessons derived from their findings to schools and youth organizations.[17]

A notable level of foresight can be seen in these early musings on nature protection in 1950s science journals and magazines. In Jin Jieliu's article, for instance, he urged Chinese readers to see the study of nature as an international issue. Jin noted Soviet participation in international nature-protection meetings and claimed that China was not paying enough attention to nature protection and the importance of understanding the conditions of environmental protection in other countries. Jin further stressed that nature protection should be seen as a means of working toward socialism.[18] The exact means by which the trinity of science, nature, and the state were supposed to benefit socialist construction remained obscure in the literature that promulgated the idea, but like most writings, it championed the necessity of the endeavor.

Zhu Kezhen, who was a strong proponent of nature-reserve establishment, articulated the connection between nature protection and national production with more authority as a leader in the CAS, but without a great deal more specificity. In his 1959 article, he wrote, "Nature and natural resources are the sources of human production. . . . The materials that human activity mainly depend on are all from nature and are governed by natural laws."[19] In his view, nature not only existed for the sake of human use, but also for the development of human society. The purpose of nature protection in Chinese communist terms was to ensure the continued presence of these important resources for the sustained support of the development of the nation and society. Zhu did not see any contradiction between nature protection and the maximization of human productivity. Scientific research on nature could enable society to learn from nature or obtain the maximum benefit from such resources as timber, oil, and water.

Although during his early career, and even through Japanese occupation, Zhu Kezhen was no communist,[20] his writings as the vice president of CAS make a strong case for communist societies' greater suitability for nature protection. Zhu Kezhen stressed that socialist societies were innately better suited to the practice of nature protection than capitalist societies. He stated, "The aim of socialist economic planning is to benefit all people and therefore has to consider issues regarding the use and improvement of nature." Along these lines, Zhu described socialist planners as "looking at the big picture and planning according to the circumstances of nature in various local areas and of nature as a whole."[21] In contrast, he believed that capitalist societies were more prone to plundering and exploiting natural resources for profit.[22] Interestingly, the articulations for state-sponsored nature conservation in the People's Republic of China were very similar to those found in writings from such capitalistic societies as Progressive-Era America, which also was concerned with the maintenance of resources for the nation's future development and had similar faith in science as a means of achieving that. Historian Samuel P. Hays noted that during the Progressive Era, conservation's "essence was rational planning to promote efficient development and the use of all natural resources. It is from the vantage point of applied science, rather than of democratic protest, that one must understand the historic role of the conservation movement [in the United States]."[23] Communist conservation and capitalist conservation shared the same basic rationale for protecting natural resources and employing scientific methods to do so. These parallels become more understandable in light of the fact that Zhu Kezhen and other contemporary leading scientists in China were educated in the United States in the early twentieth century. After becoming disillusioned with Chiang Kai-shek's Nationalist Party and refusing to retreat with him to Taiwan, Zhu Kezhen not only threw his lot in with the Chinese Communist Party, but also became an important theorist for integrating scientific ideas and communist ideology.[24] He saw that the difference between the two systems had more to do with their separate visions of what each form of development would look like.

Over the course of the first decade of the People's Republic of China, the study of nature gained official recognition as an endeavor that contributed to the building of a socialist state. Progress in nature-protection efforts, however, slowed to a virtual halt when the country was immersed in the Great Leap Forward. The intense focus on steel production and agricultural output resulted in widespread deforestation and the subjugation of broad stretches of wilderness to land reclamation efforts.[25] Much of the intellectual progress of incorporating the study of nature into the communist project was lost. In turn, so was a great deal of nature.

Turning Weasels into Tractors

Wilderness degradation was by no means the only or most catastrophic result of Mao's Great Leap Forward. This aggressive industrialization effort also produced an economic downturn and the world's worst famine in history; an estimated twenty to thirty million people starved to death.[26] From this position of desperation, the socialist state adopted a dramatically new perspective on nature and its benefit to society. Once China began to recover from the famine, the next task at hand was economic recovery for the nation. Suddenly the potential of China's wilderness seemed to open broadly, that is, as long as China's wildlife populations could be reestablished in the wake of recent devastation and allowed to stabilize.

The State Council issued a major nature-protection policy in 1962, which shifted away from ecological research and toward species preservation. The State Council directed the provinces and autonomous regions of China to establish nature reserves for the sake of wild-animal resource protection and the regulation of hunting.[27] Science was only mentioned as a means of promulgating the value of wild animal protection by including it as content "in biology classes in the schools at every level."[28] In addition to espousing the importance of wild-animal resources to the nation, the

State Council's rationale for this new focus was the potential benefit that wild-animal resources could offer the country's economy:

> Wild animals are one of our nation's greatest riches. Not only can one obtain a great amount of wild animal meat, but also numerous wild animal pelts, deer antlers, and large quantities of musk. These products are very important and useful for improving the way of life of the people and for foreign exchange. For example, of the many wild animal pelts being exported, if we only count the pelts of weasels sold between 1950 and 1961, and do not count the pelts of any other animal type, the exchange could buy 19,512 large 25-horsepower tractors.[29]

In order to achieve such payback, hunting had to be actively promoted, organized, licensed, taxed, fined, and regulated. The state no longer saw the purpose of nature reserves as enabling scientific study for the potential of future beneficial discoveries. By 1962, it viewed nature's role in their socialist society as one of immediate material benefit.

The new focus on hunting regulation partially resulted from concerns about the sustainability of these wild-animal resources. Although this interest in hunting regulation and species protection was a new priority, it was not without some precedent. A number of recent minor policy changes indicate governmental attention had recently turned toward the status of wild animals in China. The State Council assigned control over hunting to the Ministry of Forestry in 1958. Prior to this, hunting had virtually no cohesion as an industry in China. Even with this formal structural adjustment, the State Council did not have broad ambitions to actively regulate hunting. It simply recommended that the Ministry of Forestry base its hunting regulations on "local conditions and the guidance of appropriate offices within the provinces."[30] On occasion, the central government or the Ministry of Forestry attempted to regulate specific aspects of the hunting industry, such as the 1961 ban on the production and use of a particular cyanide-laced pill designed to poison animals, which had damaging effects on the quality of game pelts and endangered livestock and humans.[31] The policy passed in 1962 elaborated on these earlier and largely unenforced policies, but more important, it marked the beginning of a cohesive state agenda and the governmental embrace of the concept of a hunting industry.

The notion of a hunting industry and the protection of pelt game was rooted in academic exchanges with Soviet scientists. Soviet scientists authored many articles in Chinese journals that promoted state ownership of wild animals and the importance of compensating the state for each sale.[32] According to these articles, a central purpose in implementing the new restrictions on hunting was for China to be able to perpetuate the supply of wild animals in order to sustain hunting as a long-term industry and source of state income.

Looking back to articles from the 1950s on nature protection, evidence abounds that the basic terms of the new protection policy were strongly influenced by these earlier academic and professional exchanges. The zoological journal *Dongwuxue zazhi*, for instance, published a report by CAS zoologist Zheng Zuoxin on academic exchanges between China and the Soviet Union. As late as 1959, Zheng celebrated Soviet advancements in zoology, nature reserve establishment, and a developed hunting industry.[33] Parallels between the 1962 policies and the content of these articles are abundant, but they are not acknowledged in the policy texts due to the change in political climate following the Sino-Soviet split of 1960. Renowned Soviet zoologist Georgii Petrovich Dement'ev had considered hunting regulation to be an integral component of nature protection in terms of both preservation of species for scientific study and conserving natural resources as property of the Soviet Republic. He wrote,

> Considering that all animals are the nation's wealth in principle, it is essential that we have an established plan for the use of these resources. Hunting for sport, planned industrial use, and the hunting of animals that are harmful to agriculture, forests, and sanitation are all allowed. Aside from these, all other kinds of hunting are detrimental to nature protection.[34]

Echoes of this can be seen in the 1962 Shaanxi Provincial hunting policy written in response to the 1962 State Council national wildlife protection directive. The Shaanxi Forestry Bureau categorized hunters in a similar fashion to Dement'ev's description, including professional hunters, seasonal hunters, holiday hunters, and small-game hunters. Shaanxi forestry officials virtually quoted Dement'ev by stating, "Regardless of the type of hunting practiced, the game is always the property of the state. The hunter is to compensate the state accordingly."[35] These new national and provincial hunting regulations also promoted the hunting of animals that were, as Dement'ev categorized them, "harmful to agriculture, forests, and sanitation." The Chinese national policy of 1962 notes, "If trying to get rid of harmful wolves, jackals, and rats, it is permissible to employ annihilation hunting methods."[36] Both China's policies and these 1950s articles that referenced the USSR promoted the creation of areas where hunting was prohibited. Dement'ev also discussed the importance of the notion of "rational hunting," which was reiterated in 1962 policies. While Dement'ev did not use the term as such, he described the principles attributed to it. "It is permissible to hunt and trap animals that are abundant, but when animals have shrunken in number suddenly (like the elk) they are put back on the list of protected animals."[37] The following excerpt from China's 1962 State Council policy reflects the influence of these earlier Soviet and Chinese writings in iterations of such notions as rational use. "In areas where the resources are not yet denuded, there should also be no further threat to the continued existence of wild animal resources under the premise that a specific number be allowed for rational hunting with a plan for rational use."[38]

Several 1950s articles discussed examples of the economic potential and economic use of wildlife resources that were reflected in the new 1962 hunting regulations. In his 1959 article, Zheng Zuoxin praised the emphasis that Soviet nature-protection scientists placed on the economic production value of the target protected species.[39] Another element prominent in Soviet hunting that became a major component of China's 1962 hunting policy was the notion of *yang* (养), or rearing, including raising wild animals, stocking areas with animals, and domesticating wild animals. Both Zheng's and Dement'ev's articles hailed Soviet muskrat breeding and introduction as a success story in rearing "wild animals" for the fur trade. "Between 1927 and 1953 they [the number of introduced muskrats] grew to over 117,000 in 500 new areas."[40] Soviet scientists who later promoted a less interventionist approach eventually condemned this form of species introduction.[41] As far as Chinese policymakers in the early 1960s were concerned, however, species introduction was a sound practice worth emulating. This type of disjuncture indicated that although influence from previous cooperation with the USSR had a lasting impact on China, the scientific communities of the two countries had parted ways. Subsequent developments in Soviet science thus did not make their way into China's scientific discourses.

These theoretical influences materialized in 1962 because that year the PRC government recognized that it was at a crucial juncture. In addition to figuring out ways to integrate China's wilderness into the nation's economic structure, the government realized that the wilderness was in a precarious position, and it was genuinely concerned about the sustainability of wild animals domestically. As such, the hunting industry was part of a two-fold plan that included active protection of wild-animal resources. The State Council noted this issue obliquely by focusing on the depletion of wild animals rather than the cause. It is difficult to ascertain the extent to which the animal population had diminished during the Great Leap Forward and famine. There are no comprehensive figures in the State Council document, and such numbers probably do not exist. But the State Council enumerated a few examples and called the large-scale hunting that had occurred between 1959 and 1960 "reckless." The government viewed this reckless hunting as the main threat to the populations of a wide variety of animals in various regions of China.

> In 1960 some 6,900 wild donkeys were hunted in Qinghai. . . . That same year, military forces entered the grasslands of Inner Mongolia and killed yellow sheep, transforming a giant herd of several thousand into a small herd of just a few dozen sheep. . . . In some places, the hunting of deer and musk deer was not done in order to obtain deer antlers or the musk, but to eat the meat. In Sichuan alone, 62,000 deer and musk deer were killed in 1960.[42]

Without being explicit about a cause for this large-scale wild-animal depletion, the State Council's discussion of animals hunted for their meat point to widespread hunger as the cause. The 1962 shift from scientific experimentation and ecological study to hunting regulation therefore emerged as a reaction to famine and its ecological and economic repercussions.

The early 1960s were broadly characterized by the effort to rebuild and recover economically from the disastrous Great Leap Forward. During these years Mao receded from day-to-day management of the country and designated Liu Shaoqi as the head of state. Liu Shaoqi and his colleagues instigated broad economic reforms that, unlike Mao's Great Leap Forward slogan of "more, faster, better," focused on light industry for less hasty, but more stable economic growth.[43] Hunting was an atypical industry consisting of a scattered, inconstant workforce, no factory, and no "produced" product. In spite of these distinctions, such features as the negligible overhead associated with hunting made this industry an ideal sector to promote under Liu Shaoqi's economic program. Liu Shaoqi's strategy required the building and expansion of inexpensive industries in order to enable wide participation and maximize small profits.

The attention and effort that the Ministry of Forestry invested in transforming the disparate practice of hunting into an actual industry are strong indications of the degree to which the government genuinely saw hunting as an opportunity to contribute to national economic recovery. The Ministry of Forestry laid out detailed regulations for the management of the industry. The guiding principle of "protecting, rearing, hunting," which promoted "the strengthening of resource protection, active breeding and rearing, and rational hunting and trapping," became a policy slogan, reiterated in government documents from local to national levels during the subsequent decade.[44] The Ministry of Forestry drew up the blueprint for the development of this industry that was subsequently incorporated into the State Council's plan for the creation of nature reserves and the preservation of China's wild animals.[45]

Among the other items described in the text, the State Council outlined ways to restrict hunting practices in minute detail. Each municipality and county was responsible for protecting wild-animal resources within its boundaries. In areas where the environment or wild-animal resources had been badly depleted, hunting was prohibited. In other areas, hunting was to be restricted by specific geographic parameters, seasons, types of game, and possession of a hunting license. Methods of hunting were also closely regulated. Poison, dynamite pits, mechanized pursuit, illuminated hunting, and other dangerous or damaging methods all were prohibited.[46]

The new hunting industry was managed by the subordinate offices of the Ministry of Forestry. The State Council reiterated that all wild animals were national property and their preservation was the duty and responsibility of all citizens of China.[47] Forest hunting areas were placed under the jurisdiction of an appointed person who was in charge of collecting fees or a percentage of the hunt. Entrance into hunting areas was by license and local government approval only. Hunters were expected to sell pelts, medicine, and other such wild-animal products to the state in collaboration with the Ministry of Commerce.[48] The Ministry of Forestry granted district officers a great deal of discretion. They could assess fees according to individual circumstances. Where existing rates might compromise the livelihood of the individual hunter, district and county officers were encouraged to reduce the fees. The State Council also maintained a few concessions, such as the allowance for unregulated hunting of small meat animals.[49] This clearly was a nod to the persistent need for food. In this way, the modern state of the PRC demonstrated that it was still capable of observing some forms of a "moral economy."[50]

Under the new system, hunters could hire themselves out to work as professional laborers for the Ministry of Forestry office that regulated a specific forest area. Their earnings were to be "no less than that of other Ministry of Forestry laborer salaries."[51] By creating paid positions of state hunters, the new policy transformed widely dispersed people into a group resembling a more typical work unit.

The government's interest in the economic potential of China's wilderness was not an abstract policy. By coordinating China's top legislative body with local bureaucrats in China's most remote mountainous hinterlands, a new industry was created and the reach of the state dramatically extended. At the same time, local bureaucrats became heavily involved in this new project and put their own stamp on the shape of its development. The relevance of nature to socialism was elevated, and the hunting of wild animals was regulated because of the economic need for their sustained use. According to government documents, the years of the famine had given rise to natural resource depletion on a scale that was previously inconceivable. During the Great Leap Forward, industrial growth had become more important than human survival. In the wake of the famine, China found itself needing to explore diverse methods of recovery. Just as peasants had increasingly turned to wild animals for sustenance, the national government found industrial growth opportunities in this same highly marginal resource in its geographic periphery. Conversely, as much as the hunting regulations of 1962 were a response to the famine and reflected extreme economic need, they also indicated that China was on its way to recovery. By 1962, with surreptitious grain importation and the recovery of crop harvests, China was finally in a position to restrict hunting for meat.[52] In the end, the wild-animal protection policy was simultaneously a response to economic crisis and a post-famine luxury. Not all nature, however, was equally valuable, and China's government had to assess the relative worth of its different types of wildlife vis-à-vis state needs.

"Precious and Rare"

Authors of the 1962 nature-protection policy included a special category of restriction among the many other hunting regulations embedded in the policy—the prohibition against hunting animals that were considered to be "precious and rare" or *zhengui xiyou* 珍贵稀有. Government officials and scientists alike widely used this term, also often abbreviated as *zhenxi* 珍稀 in their writings during the 1950s and 1960s, to describe physically distinctive animals that were endemic to China and, according to Chinese forest surveys or the impressions of local residents, few in number. Wildlife surveys conducted during the 1950s were one source of information that may have indicated how rare an animal was at that time, but no species-specific surveys had been conducted. The Ministry of Forestry formulated a wild-animal protection policy in 1959, which prohibited the hunting, trapping, or export of nine precious species, including the giant panda and the golden-haired monkey.[53]

Concern for and interest in these special species can also be traced to 1950s scientific articles. In Jin Jieliu's 1957 article, "Guanyu 'ziran baohu'" (关于 "自然保护" Concerning "nature protection"), he argued that the desire to prevent the extinction of rare animals fundamentally was a concern for the loss of unknown and potentially useful knowledge. This sentiment aligned with the valuation of nature preservation that fed the creation of China's first reserve and fit well with the current intellectual atmosphere. Jin's interest in nature protection, however, expanded beyond the scientific focus of projects like the Dinghu Shan Nature Reserve. Jin Jieliu believed that a rare indigenous species could serve the purpose of a natural monument—a tribute to China's natural wealth. Jin thought that a species played such a role most effectively if it was unique to a particular region and country, specifically giving the example of the giant panda.[54] Seeing the giant panda and other rare animals as monuments introduced a role for rare natural phenomena as objects of symbolic value. This may offer some insight into the reasoning behind the dedication to precious-species protection. A sense of nationalistic pride in China's ecological and biological uniqueness is evident in this notion of a living national monument.

The purpose of prohibitions against hunting "precious and rare" animals differed from those protecting more common animals. The point of protecting the former was not to replenish a game supply, but to preserve unique and disappearing species. The 1962 policy listed nineteen animals, the giant panda first among them.[55] Hunting these animals was prohibited. No accommodation was made for future restricted hunting dependent upon species stability. The value of "precious and rare" animals was not based on their potential as future sources of expensive pelts or medicinal products. Indeed, their value for the state and its ministries remained quite nebulous in government documents. Policy texts described these special fauna as significant because they were "more than simply economically valuable or important for scientific interest."[56] In spite of this vaguely expressed value, "precious and rare" wild animals became the centerpieces around which new nature reserves were to be constructed.

> For these animals, hunting is strictly prohibited. Moreover, nature protection reserves will be established in the primary habitat of these animals and in their breeding areas in order to better protect them.[57]

Although the apparent impetus for the 1962 policy was to aid economic recovery, the creation of the systematic protection of these particular species did not offer any clear economic benefit at the time or even potentially in the future.

The policy demonstrated that concern about the original nine animals became more acute and that there was reason to be concerned about a broader spectrum of species. There is little evidence that the 1959 precious-species protection policy had been widely disseminated, but even if it had, it by no means spared these animals from depletion. Precious species had been just as vulnerable to "reckless hunting" as other animals when the Chinese people suffered from widespread hunger. Giant pandas reputedly had poor-tasting meat. Their coarse pelts were not highly sought after.[58] The giant panda was neither commonly hunted as game nor in self-defense.[59] Locals viewed golden-haired monkeys as too human-like to eat.[60] Such preferences became irrelevant in the face of starvation. As was the case with many large meat animals, the number of giant pandas, golden-haired monkeys, and other precious species dropped dramatically during the years of famine.

> In Shaanxi province, in the Qinlin area, about one hundred golden-haired monkeys were killed en masse in 1959. Since 1960 there remain just a few groups of elephants in Yunnan province. The hunting of Siberian tigers, sika deer, giant pandas, peacocks, and other precious animals continued during these difficult years. In Pingwu county, located in northern Sichuan, forty pandas were killed at that time.[61]

Given that people were eating dirt, bark, and purportedly other humans, the killing of nontraditional meat animals would not be unexpected, but as a result, most of China's "precious and rare" species became even rarer and more precious.

The concept of precious species protected from any hunting whatsoever also inspired *active* protective hunting. This was expressed in language that resembled a code of ethics for anyone engaged in hunting. "Anyone who enters the mountains to hunt must be careful to prevent forest fires and to protect precious and rare species and the wellbeing of beneficial birds and animals. Moreover hunters should work to annihilate any [animal] that is harmful to beneficial birds and animals."[62] Hunters therefore were not only encountering restrictions on what, when, and how to hunt, but were also expected to exert energy and time to conduct "protective" hunting. This expectation was framed as a responsibility that came with the privilege of hunting in the forest. These regulations transformed the categorization of animals in addition to differentiating beneficial from harmful animals. The destruction of harmful animals was akin to predator control, which was a prominent component of Soviet "nature protection" and a popular concept in western countries.[63]

The distinction between beneficial and harmful animals was taken up in all of the related policy documents and the majority of related journal articles. From the State Council down, great efforts were made to distinguish the harmful from the good, though the lines between beneficial and harmful animals were often quite gray. In general, the former were considered those that had something to offer humanity—fur, meat, or medicine—and the latter were generally those that threatened livestock, humans, or agriculture. Because of concern over livestock and grain, wolves and rats were mentioned most frequently as typical harmful animals. One article inconclusively debated the categories of "harmful" and "beneficial" with regard to the wild rabbit.[64]

> From the perspective of pelt animals, the wild rabbit is an economically valuable animal worth hunting and therefore beneficial to humanity. However, it threatens agricultural crops and influences agricultural output. It therefore is undoubtedly harmful to humanity.[65]

This sincere concern over how to categorize the wild rabbit was associated with a new conceptualization of nature that had ramifications in the form of regulations and behavioral expectations. As a hunter, one's duty to shoot or protect a specific animal depended on an understanding of this categorization. If the rabbit was considered beneficial, the hunter's duty was to protect it and help foster its population growth for pelt trade through regulated hunting. As such, it was also game that had to be sold to the state or, at the very least, taxed. If the rabbit was labeled a harmful pest, on the other hand, the hunter's duty was to annihilate it, and there were absolutely no restrictions concerning its hunt.[66] The wild rabbit could also conceivably fall into a third category, that of small game animals that were not taxed or regulated. The conundrum over an animal as seemingly insignificant as a common rabbit exemplifies the extent to which even a very basic classification system can be quite unclear, which made navigating the forests and wild spaces of China a particularly tricky endeavor for the new hunter, especially when serious penalties fell upon those who did not adhere to the new policies.

The implementation of these various policies proved to have some positive effects. A zoologist in CAS named Zhu Jing commented on the restoration of China's wildlife because of the three pillars of the hunting industry, "protecting, rearing, and hunting."[67] He boasted of China's success in wildlife protection, noting a twenty percent increase in the number of red deer (wapiti) from 1960 to 1963. Zhu also cited the increased success of the hunting industry with great enthusiasm, noting "an annual average of twenty million pelts 'produced' by China's hunting industry."[68] Zhu Jing's excitement about the positive impact of the hunting industry on the national economy and socialist development reflects his view that wild animals were natural resources with clear material value.

In the realm of "rearing" wild fauna, he noted the successful captive birth of a giant panda in 1963 with the caveat that "even though the giant panda is not an economically valuable animal, this is an impressive accomplishment."[69] This quote offers clear evidence that the panda's value at that time was not financial, but as a natural treasure. Zhu's writings further demonstrate the ways that such categories as "economically valuable" and "precious" were supposed to take effect with the implementation of the policy. What animals in these categories shared was that they were natural treasures of China. Preserving "precious" animals protected China's wealth in natural beauty and biological diversity. Protecting "economically valuable" animals sustained China's ability to grow its wealth.

Desperate for broad-scale, steady economic growth in 1962, the Chinese state reexamined the wilderness of China and decided that it had more to offer than potential scientific insight. Villagers who previously supplemented field harvests with game meat became mountain laborers with spatial and temporal restrictions along with new duties, new game, and new ethics. The state became a buyer and overseer, and as such, simultaneously responded to the dual problems of economic crisis and expansive wildlife depletion. Amidst all of this, a class of animal in potential danger of extinction, in need of protection, and without much to offer except for an innate uniqueness was converted into a natural treasure. The 1962 policy was genuinely transformative. It formally engaged China's natural areas through economic integration and ushered in a new view of the value of China's nature. This era also enabled the expression of pride in China's natural uniqueness, particularly as manifested in individual species, and a sense of duty to preserve China's natural treasures. Endemic and rare wild-animal species thus carved out a niche for themselves in China's economically strained socialist experiment, that of highly valued natural resources with very little material value, at least for the time being.

3

The Winding Road to Wanglang

CREATING A PANDA RESERVE

THE FIRST PHYSICAL manifestation of the 1962 nature reserve policy was the creation of a small, remote reserve in Northern Sichuan Province for the protection of giant pandas and other local species. Just months after the State Council distributed the directive to all provincial-level forestry departments, the Sichuan Department of Forestry tapped Pingwu County in the distant northern part of the province as a potential site for building a nature reserve for panda protection.[1] Three men who were instrumental in writing the directive were biologists from Sichuan Province, which perhaps explains why province officials responded so quickly to implement it.[2] The Sichuan Department of Forestry wrote a plan to set up several nature reserves there, reflecting its strong support for the goals of the central government policy. Both the original directive and the Sichuan Province plan put an emphasis on wild-animal protection, including hunting regulation and a concern for unique and rare species.[3] The provincial government saw potential value in Sichuan's wildlife resources not only because they were varied and abundant, but also because the native flora and fauna were spectacular and unique. The Sichuan forestry officials stated that Sichuan's participation in this campaign was vital: "Our provincial wild animal resources are extremely rich and abundant, not to mention that we have many world famous, 'precious and rare' birds and animals which comprise a significant component of our nation's great sources of wealth."[4] Of the nineteen animals targeted for national protection, nine are indigenous to Sichuan.[5] Not only did the province's Department of Forestry argue that nature-protection work was particularly applicable there but also that, consequently, Sichuan was particularly valuable to the nation.

Zhong Zhaomin, a diligent and responsible man working in the Pingwu County forestry office, was called upon to take up this project. Early one spring morning, he, three experienced local hunters, and a pack mule began walking along the horse trail toward the Wangbachu forest station near the Wanglang forest area in the northwest corner of the county. After they had made their four-day trek, they set up a base camp at the primitive forest station and began assessing the land and its resources.[6] In the extremely bitter conditions of the northern Sichuan mountains, this small force recorded the flora and fauna, topography, and condition of the bamboo in the area.[7] This final factor was of particular concern because they were seeking the most suitable area in the county in which to create a giant panda protection reserve. Once they finished surveying the Wanglang forest area, they ventured to an even more remote section of the county and conducted a comparative survey of the Si'er forest area.[8] This venture, however, involved much more than a dutiful response to a government order. The decision to place the reserve in the Wanglang forest area required the integration of national, provincial, county, and local concerns. It involved the transformation of a national government policy into the tangible reality of the Wanglang Nature Reserve. Zhong Zhaomin not only attended to local concerns but also created a productive relationship with the national government by creating a space for national-local interaction, which was vital to China's early wild-panda preservation work.

From the Top Down

The project of creating reserves for precious species protection exemplified the state's assertion of authority and ownership over China's wild lands as well as specific animal resources within them. The central government's program promoted state reclamation, regulation, and rejuvenation of the country's wildlife. Reserves could advance this program as a designated space for strict control of hunting and the protection of animals that the State Council specifically named as valuable to the nation irrespective of their local significance. By making a list of precious species, the central government reiterated that each was national property and thus subject to government regulation and protection. By placing the implementation of this policy and the creation of nature reserves in the hands of provincial and county governments, however, the central government created a structure for the national policy to acquire local characteristics. The actual responsibility for creating the reserves fell to the lower levels of the bureaucratic structure. With this transfer of authority, the purpose and goals of nature reserves shifted from the broader project of establishing a socialist state to concerns more specific to the province and county. At the same time, local and provincial participants saw their implementation of this policy as a means of working for the nation.

The Sichuan provincial forestry office responded to the national initiative with a reiteration of the State Council's main points and suggestions for how to "combine the essence of the State Council's guidelines with the practical circumstances of our province."[9] Similar to the central government, the Sichuan office placed the economy at the forefront of its directive for wild animal protection.[10] Explicitly prioritizing localized economic growth, the provincial forestry department promoted the implementation of these policies as a means to benefit provincial mountain areas and better the "everyday lives" of the people in Sichuan.[11] It strongly encouraged trade with outside communities as a means of realizing the economic potential of its wild-animal resources.[12]

The province decided to focus its efforts on creating reserves for the first animal listed in the State Council document, the giant panda. In addition to Pingwu, three other counties—Wenchuan, Nanping, and Tianquan—were assigned the task of creating panda preserves.[13] The provincial office also proposed but did not establish a musk deer reserve in Muli County.[14] The giant panda shares its habitat with other "precious and rare" species that are also protected by the state. Two of these, the golden-haired snub-nosed monkey and the takin (a large ungulate), were named as additional protected species in the panda reserves. In the provincial report, however, the giant panda was given top billing in each of these reserves.

The provincial document also introduced the notion of a nature reserve as an important and unique environment in which to conduct scientific research. The introduction of science into policy rhetoric at the provincial level is notable because this introduction was not mentioned in the State Council document. Interest in the scientific study of nature in fact reflected the nature-protection efforts in China during the 1950s and was the main impetus for the Chinese Academy of Science in establishing the PRC's first nature reserve, Dinghu Shan, in Guangdong Province in 1956. Scientific research was by no means the main focus of the 1963 Sichuan directive, but its inclusion among the enumerated goals of nature-protection work influenced the design and implementation of the project at the county level.[15]

Pingwu, a little-known county where pandas had been sighted, is situated 305 kilometers north of Chengdu, the provincial capital.[16] The government's knowledge of pandas' presence in Pingwu was originally drawn from incidental reports and observations made during the extensive forest surveys of the 1950s.[17] Subsequently, eleven of the thirty-five giant pandas captured for domestic exhibit by 1963 were caught there.[18] Giant pandas have a general preference for high altitude mountains in cold climates, which made Pingwu an ideal environment. The Min Shan mountain range that runs through the county features peaks that reach heights of over 5,000 meters, and the climate is cool, with an average annual temperature of 14 degrees Celsius (57.2 degrees Fahrenheit), which drops significantly as the altitude increases.[19] In the area where the Wanglang Nature Reserve was later established, the altitude ranges from 2,800 to 4,000 meters, and the average annual temperature in 1965 was 9 degrees Celsius (48.2 degrees Fahrenheit).[20] Pingwu's climate is conducive to the growth of an alpine forest and, what is more important from the perspective of the giant panda, a thick growth of arrow bamboo.[21]

Of the five Sichuan counties designated to set aside land for establishing a nature reserve, Pingwu was the first to act. Just as Sichuan's keen interest in establishing nature reserves was linked to the personal affiliations of national officials, Pingwu's prompt and thorough response was due to some key individuals.[22] Former deputy director of the Wanglang Reserve, Feng Yunwu, attributed the success of Pingwu's pioneering efforts to Wanglang Reserve's founder, Zhong Zhaomin. A modest man, Zhong credited his superiors in the county and provincial governments for moving the nature reserve project forward. He noted that Hu Tieqing, head of the provincial nature-protection office, was instrumental in promoting the project at the provincial level. In turn, Zhong Zhaomin praised the head of the Pingwu County People's Committee, Zhang Zaikun, who "placed a tremendous amount of emphasis on this work."[23] Zhang had selected Zhong Zhaomin, transferred him from the county public affairs office, and put him in charge of what turned out to be the creation of China's first panda reserve.[24] Having studied biology in Mianyang, a city halfway between Pingwu and Chengdu, and forestry at the Beijing Academy of Forestry, Zhong Zhaomin was one of the most highly trained staff in the county.[25]

Zhong's first challenge in creating Sichuan's first nature reserve was navigating the topography of Pingwu County. Although his work and education gave him the experience and knowledge necessary for such a task, he knew that he needed help to survey the land effectively. He found three able assistants, Geng Shengwa, Cai Li, and Chai Ningzhu, hunters who were also skilled in herb collecting and were intimately familiar with the region's mountainous wilderness.[26] With their assistance, Zhong set out in search of an area to protect for its rare animal inhabitants.

The characteristics that made Pingwu County appealing to pandas made surveying it difficult for humans. High elevation, steep slopes, cold temperatures, and thick undergrowth slowed the progress of Zhong's team.[27] The physical landscape of the Min Shan range imposed its own restrictions on the placement of the new reserve. As you can see in Figure 3.1, while the thick bamboo undergrowth was difficult to navigate, even more challenging were the steep mountainsides that commonly sloped toward the sky at 40-degree climbs and sometimes angled upward even more sharply.[28]

After the arduous surveying work was complete, Zhong Zhaomin wrote a concise recommendation to place the reserve in the Wanglang forest area rather than in the Si'er forest area, as there was a need to protect the virgin forest and ancient topography of the site. Although the previous directives from the provincial and national levels both stated that the habitat in the areas of the wild animal reserves should be protected, Zhong was the first to recommend targeting specific rare trees for protection.

In addition to expressing a special concern for Pingwu's ancient forest, Zhong Zhaomin wanted to preserve the ancient forests of the area. Zhong noted that because there was little evidence of environmental degradation around this particular area of forest, it appeared to be a "natural nature reserve."[29] While topography was not an issue of concern in the national or provincial directives on protecting wild animal resources, a precedent for topography preservation could be found in another provincial-level document issued on the implementation of nature-reserve work just a few months earlier. In that document, the Sichuan office remarkably attributed the rapid changes in the appearance of the natural world to the "development of socialist projects."[30] Even more striking was the expressed desire "to preserve the natural world's ancient appearance" not only for the sake of preserving natural resources and conducting scientific studies, but also so that "future generations will be able to see the ancient appearance of the natural world."[31] This aesthetic goal transcends the pragmatic tone and ideology most commonly expressed in government documents from this period. Zhong Zhaomin subsequently was able to categorize such factors as the appearance of the landscape as a relevant category for analyzing preservation needs of the county and province. As such, provincial and county policy recommendations expanded the scope of priorities for national nature protection.

FIGURE 3.1 This slope in the Wanglang National Nature Reserve exemplifies the terrain of the area. Photograph by E. Elena Songster, Wanglang Nature Reserve, Pingwu County, Sichuan, 2002.

In his county report, Zhong Zhaomin also addressed the issue of efficiency of space in nature protection. Because the area of Wanglang contained "the greatest number of wild fauna and wild flora in the county," it was a logical place to set up a nature reserve. Zhong reasoned that by drawing the reserve boundaries around the area of Wanglang, the county could protect the greatest number of animals and plants with the least amount of land investment. After conducting a number of surveys and interviews, Zhong determined that at least one hundred mammal species lived in the 12,000 hectares recommended for this reserve.[32] Although the giant panda was to be the target protected species, all of the other animals that shared its habitat, not just the "precious and rare" and "economically valuable," were protected within the boundaries of the reserve.

Zhong indicated in his report that scientific research conducted in the Wanglang reserve would provide an important service to the nation. Early projects could include tree cataloging, extensive surveys of plant and animal life, and the promotion of wild animal reproduction.[33] The impetus behind the central government initiating nature reserves was simultaneously preserved and transformed as the implementation process moved through the chain of command. At each lower level, the official documents contained an increasing number of added concerns. The authors of these government policies began with a simple vision that focused on hunting regulations and wild animal protection. At the provincial level, this grew to include more specialized economic growth, trade, scientific research, and aesthetic beauty. The national government routinely depended on the greater familiarity of local officials with the areas in question and the broader understanding of local issues that might affect the implementation and efficacy of any given national policy. Many examples were not harmonious—some reflect too much control from above, others indicate that some local governments should have been more tightly controlled. Indeed, there are many documented examples of dramatic corruption at the local level in which this autonomy proved to be detrimental to local populations.[34] The creation of the Wanglang reserve offers an example of a constructive interaction between the vast People's Republic of China and a small county, largely due to the involvement of an individual with insight who cared about the project.

From the Ground Up

Among the specific features that made Wanglang a preferable site for China's first panda reserve was accessibility. Although Wanglang was a long journey by foot from the county seat, as Zhong Zhaomin explained, "Si'er was even more difficult to get to than Wanglang; it therefore was in less danger of being denuded."[35] Even though protecting wild animals in the reserve regardless of their protected status required inactivity in the reserve, the monitoring and control of forbidden activity required that the reserve not be completely inaccessible. In order for scientific research to be conducted, people also had to be able to travel to and from the reserve. Wanglang's greater accessibility would thus better facilitate the county forestry office's work.

Protecting the panda habitat in Wanglang also required particular attention to bamboo, the giant panda's primary food source. Zhong's survey team reported, "Arrow bamboo grew thickly in the shelter of the alpine forests."[36] Zhong's attentiveness to the bamboo stands, however, was much more sophisticated than this simple observation implied. Zhong, pictured in Figure 3.2, researched the status and longevity of the bamboo supply in each forest area beyond what was simply observable on the ground. He understood the natural life cycle of bamboo and the periodic die-off that accompanies flowering and seeding. The length of this cycle varies with each species of bamboo and can be affected by environmental factors, but cycles tend to be quite long, varying from forty to eighty years.[37] Zhong learned that the last time that bamboo flowering had been recorded for the Wanglang area was 1933; in Si'er, bamboo had flowered just a few years prior to his survey, in 1958. Although he did not elaborate on the history and status of the bamboo stands in his report, they were significant

FIGURE 3.2 Zhong Zhaomin, founder of Wanglang Nature Reserve. Photograph by E. Elena Songster, Pingwu City, Pingwu County, Sichuan, PRC, 2002.

factors in his assessment of the Wanglang area's suitability to support a greater number of giant pandas.[38]

The final factor in the selection of Wanglang connected to its distinctive geographical traits. The Wanglang forest area constituted the entire northwest tip of the county, with two of the reserve borders sharing two county borders. Such an orientation could enable a nice congruence in the protection of county and national natural resources. Water was a significant component of the physical landscape. The origin of the Baima River, the main water source for the county of Pingwu, was located in the Wanglang forest area. Protecting it would benefit both the pandas and the people. Even while Zhong designated an area of nature outside of local communities and industry for the reserve, he saw the natural and human worlds as an integrated whole. Placing the reserve around the water source would allow this valuable natural feature to be monitored and protected for wildlife in the reserve and the people of Pingwu. Wanglang can be located on the map in Figure 3.3 in the context of the panda's current range and other major reserves. Later, the construction of a small dam on the Baima River enabled the reserve to generate its own electricity and to become certified

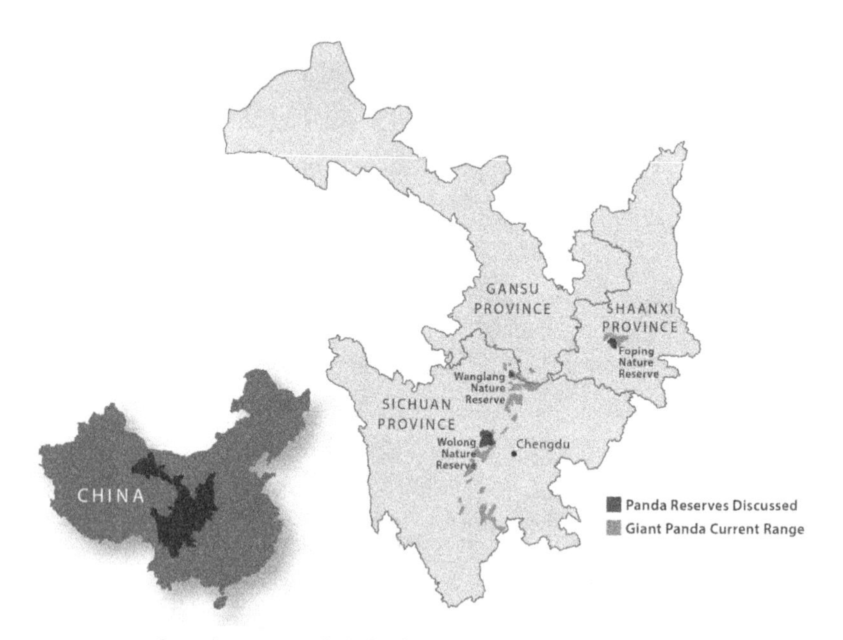

FIGURE 3.3 China has expanded the land dedicated specifically as giant panda reserves. Three reserves that have been the focus of discussion in this book and other major research projects are highlighted on this map. Map created by David Medieros, 2015.

as an eco-tourist site. Although Zhong did not foresee this benefit, he had enough foresight to recognize the river's value.

Another component of the landscape made the placement of the Wanglang Nature Reserve a matter of great urgency. The area of Wanglang contained a virgin forest and rare trees worthy of protection. As Zhong observed in his survey, part of the Wanglang forest area had been logged during the 1950s. This physical scar in the landscape, combined with the continued activity of the logging company in the Baima River valley below the potential reserve site, provided strong evidence that the rest of the virgin forest and the rare trees were under genuine and immediate threat.[39]

The recorded history of logging in Pingwu County dates back to the Ming dynasty (1368–1644). During the mid-1400s, an imperial edict called for the harvesting of cedar trees in Pingwu to repair the temples that exclusively used that kind of wood.[40] Loggers returned to Pingwu during the mid-1500s and the early 1700s. Approximately twenty private logging companies were established in the area during the Qing dynasty (1644–1911). As the Republican period (1911–1949) progressed, however, lumber could not compete with the opium trade for profit. The lumber trade plummeted, leaving few remnants of the industry when the People's Republic was established.[41] When the Communist Party finally put a stop to opium production and trade, it reestablished logging in Pingwu County.[42] The most widespread logging occurred during the 1950s. At that time, logging there was government owned and operated and thus served the nation. As one observer noted, "The state sent in loggers and took out trees; it had little to do with the county, except that the trees were found there."[43] Lumber companies such as the Northern Sichuan Forest Trade Office [*Chuanbei sen gong ju*川北森工局], later called the Northern Sichuan Logging Company [*Chuanbei famu gongsi* 川北伐木公司], began their operations with a few hundred workers, which grew to a few thousand workers, the vast majority of whom were from outside the county.[44] The increased need for timber was attributed to the rapidly expanding railroad.[45] In 1952, beginning in the county's southern district of Daying, the Northern Sichuan Lumber Company clear-cut all of the trees from halfway up the mountain down to the Pingtong River.[46] That year the company conducted forest surveys in the other parts of the county. By the next year it had already set up transportation lines for the lumber from the area of Wanglang in the northwestern corner of the county down to the county seat, ceased its operations in the southern part of the county, and moved more people and equipment to the Baima River valley, just below the area that became the Wanglang Nature Reserve.[47] The logging company was stripping the county of its trees, offering few jobs to locals, and giving the county little in exchange for this resource. The state did not pay the county for the wood, and the county even had to buy the wood from the logging company.[48]

This flurry of logging activity had a devastating impact on local communities. The Baima Tibetans, a local ethnic minority group who were the main inhabitants of the river valley, had strict traditional regulations about tree felling. They used trees for constructing their own houses and other essential needs, but they were not permitted to fell trees randomly. According to one Baima man, nobody in the community was allowed to cut trees on the mountain slopes, especially above the villages or near graves. When asked, "What if someone did cut down such trees?" he gestured with his index finger and said very emphatically, "The community would scold and punish him. It was very serious!"[49] His expression softened as he looked at the light green shrubs surrounding his home. "All of the hills around this area used to be covered with primary forest of pine and other evergreens. . . . Most of the trees were cut down during the 1950s."[50] Another Baima villager noted that the punishment for felling trees on graves or mountain slopes, in addition to community scolding and humiliation, included a requirement that "one would have to share her or his crops, cows, or opium with the community."[51] The intrusion of the lumber company changed these dynamics. She said,

> They felled trees at will, all over the mountain slopes and everywhere, without any concern for or interest in local cultural restrictions. As a result the Baima too began to cut down trees on the slopes, abandoning the traditional views. If they did not cut down the trees, the Han Chinese would. The trees would all be gone and we would not even have any wood to show for it.[52]

The county gazetteer did not record the Baima response to the dramatic and sudden logging activity around their communities, but it did note that the logging fervor continued during the late 1950s, most likely linked to Chairman Mao's Great Leap Forward campaign. As visible in Figure 3.4, some hillsides near the Wanglang reserve are still recovering from earlier extensive logging. According to the gazetteer, Mianyang prefecture, which had jurisdiction over several counties, including Pingwu, encouraged

FIGURE 3.4 The dramatic deforestation that occurred on the hills near the Wanglang Reserve is evident in this photograph. Photograph by E. Elena Songster, Pingwu County, Sichuan, PRC, 2002.

massive felling, but "only a fraction was transported and a great deal of felled timber wasted away on the hillsides."[53]

In his report, Zhong Zhaomin emphasized the importance of protecting the virgin forest, ancient landscape, and rare trees. While most of the virgin forest in the Wanglang area had not yet been cut when he surveyed the area, the section that had been felled offered Zhong a glimpse of what could occur if he did not make an effort to save the trees in the name of the pandas. The threat of deforestation did not seem to loom over Si'er in the same way. Thus the protection of pandas facilitated the protection of rivers, forests, and myriad other wildlife that shared its habitat.

The People of Pingwu

The communities of the Baima Tibetans [白马藏族] situated very close to Wanglang also affected the placement of the Wanglang Nature Reserve. However, they were mentioned only once in the various documents associated with this project. The first line of Zhong Zhaomin's explanatory report notes, "The Wanglang Nature Reserve is situated in the northwest corner of the Pingwu Baima commune."[54] The presence of the Baima people influenced the boundaries of the reserve—to be drawn outside of their villages and agricultural areas—but it is still difficult to tell from official documents the degree of their influence.

The Baima people's primary economic activity was agriculture, followed by herding. The climate and landscape were not highly conducive to farming, allowing only one crop a year of barley, hemp, corn, buckwheat, and potatoes.[55] Because the reserve existed outside the Baima people's agricultural area, Zhong felt that its presence would not encroach upon their activities too much. Herding, however, required more space. Most households were able to raise sheep and goats and a stock of chickens and pigs but had to find places to pasture their livestock. The Baima people had seasonally grazed cows, sheep, and goats in the reserve area and hunted animals and collected herbs there.[56] The Baima felt some resentment about this loss of pastureland and access to the medicinal herbs because of the placement of the nature reserve. One Baima villager noted, "There were rumors that families of the reserve officials went to the reserve to pick herbs, and such, but I'm sure such talk was purely rumor."[57] Zhong was aware that the Baima had less space for herding and herb collecting, but he did not see the reduction in land as an insurmountable problem and generally did not consider the Baima to be very threatening to the reserve and its mission to protect giant pandas.

Protecting the Pandas from the People

The government document that first addressed the relationship between the Baima people and the Wanglang Nature Reserve was a county status report that was issued to the Sichuan provincial authorities in 1965, just a few months after the Wanglang Reserve was established. Although the report did not engage the classification and identity debate of the Baima by using the commune as an identifying marker, the authors were able to distinguish their subjects from various "other" Tibetan branches.[58] This report was issued as a response to the Sichuan provincial directive that the county implement the 1962 hunting regulations and educate its residents on the new wildlife protection policies. The county government not only accepted this responsibility to enforce national policy but also used it to integrate the Baima people into the county and nation by requiring them to change their perception of the nature that surrounded them and follow parameters for what species to protect.

The Sichuan Provincial Forestry Department had expressed particular concern about a recent spate of what it termed "reckless hunting."[59] It wanted to ensure that the counties were aware of the severity of the recent wildlife depletion and were doing something about it. The terms "reckless" and "rational" refer to principles of hunting and nature protection that were developed in conjunction with Soviet biologists during the 1950s, but how animals fit into these categories depended on the type of game. For this reason, educating local residents about the government's classification of animals was crucial to enforcing policy and integrating local citizens into the national project.

Before the 1963 provincial directive to set up reserves, the county government asserted that it was not just unaware of the earlier national policies written for the protection of specific animals, but also that it maintained the perspective to "emphasize hunting, disregard protection" [着重猎取, 轻视保护].[60] As its first step in altering this approach to local natural resources, the Pingwu County Forestry Bureau conducted a survey in the Baima community about hunting practices and precious-species game, which revealed that the Baima people hunted and killed 40 giant pandas and 52 takin in Pingwu County between January and October of 1963.[61] No explanation is given for this figure, which is striking considering the virtual uselessness of the giant panda as game and the traditional Baima taboos against killing giant pandas. This hunting purportedly took place as people were recovering from the famine. Regardless of the impetus for the hunting, the county had to correct public behavior to demonstrate to the higher levels of government that it was fulfilling its responsibility. In failing to prevent the large number of precious animal deaths, the county of Pingwu felt it had failed in its duty to the country. After citing the statistics from the Baima survey, county officials commented that it was a "great loss to our nation."[62] In 1963, the act of killing giant pandas was illegal but did not carry the same weight as it would today, or in the 1980s, when some people were sentenced to death for the crime.[63] Viewing itself as the steward and caretaker of the nation's precious species, the county accepted responsibility for enforcing the new principles.

The Sichuan provincial forestry office had proposed that the county employ a wildlife and hunting management system that stressed education, including the posting of announcements enumerating restrictions and establishing hunting regulations. To publicize the regulations, it recommended use of "newspapers, journals, radio stations, posters, blackboard announcements, and the opening and broadening of informational activities." All of these efforts were expected to "expand the ability of the masses to understand the importance and significance of protection and the rational use of wild animal resources."[64] The Sichuan provincial government issued a follow-up directive the next year that specifically called on its own county governments to promote and enforce the new State Council wildlife protection policy.[65] This indicated a provincial-level commitment to enforce national policy through the county-level government. It was an example of how the vertical structure of government was supposed to function with regard to policy implementation.

Baima interviewees remember these reeducation efforts. One noted, "We didn't know that the giant panda was important to the nation before the government said, 'Protect the giant panda.' "[66] Provincial authorities did not state specific punishments for breaches of policy, and each county was responsible for assessing the gravity and punishment of any such crime.[67] This county report detailed the steps that Pingwu County was taking and punishments that the county was dictating in an effort to enforce hunting regulation and panda protection.[68] Through a series of examples the report traced the transformation of hunting behavior of local Baima citizens to demonstrate to the provincial forestry office that the county was taking the new wildlife protection and hunting regulations seriously and was effectively implementing them.

The authors of the report asserted that the Baima, based on their survey, were responsible for the majority of offenses against the new hunting regulations associated with the creation of the Wanglang reserve. Specific offenses were then highlighted. For example:

> In August of 1963 Kang Shengwa, a Tibetan member of the large Baima combat forces, trapped and killed two giant pandas. A general meeting of all commune members was called. The gathered concluded on the spot that in addition to owing a certain amount for the economic and political loss of the two trapped pandas, Kang would not be allowed to keep the hunted goods.[69]

The description of Kang Shengwa as a combat soldier conjures up images of a militaristic and aggressive man capable of single-handedly hunting such a large animal. Such typing of ethnic minorities was not atypical in Han-generated writings.[70] In the context of this report, these associations also served to magnify the county government's accomplishment in enforcing the new hunting policy—even a soldier and hunter from a remote minority group could not escape the consequences of breaking the new laws. A more significant indicator of the bureau's achievement in this story was the group meeting of commune members to censure Kang Shengwa for killing two giant pandas. That his action was purportedly recognized by his own community as wrong and punishable was deliberately included to illustrate that the county forestry bureau was successful in educating the local minority nationalities about the new wildlife protection policies. The villagers' group judgment ostensibly indicated their acceptance of the new policies, or at least their acknowledgment of what was expected of them. Observing this event through county forest bureau documents makes the assessment of the community participants' reactions impossible. Even if the local communities were not enthusiastically supportive, the documents indicated that the new policies had a notable impact on their behavior and practices.

While Kang Shengwa's case was the first cited example of punitive enforcement in the county, he was not the only offender worthy of note. The following year, eight people killed a takin. For this act, the county bureau of forestry and the home commune issued the eight hunters harsh, but unspecified punishments. In addition, each had to pay seven yuan, an amount that could purchase a month's worth of food in an urban canteen at the time.[71] This amount, worth much more in the countryside, was therefore a very severe penalty for a rural peasant. The message that wildlife protection was being taken seriously in Pingwu County was thus passed down through local society and up through the bureaucratic channels.

Pingwu County's 1965 report repeatedly mentioned the Baima as offenders against the hunting regulations, which gave the impression that they were the only hunters in the county. There was just one non-Baima hunter mentioned in the report, but only as an example of someone adhering to the new regulations.

> In 1964 a seasonal worker from a lumber mill under the jurisdiction of Mianyang district trapped a giant panda. He reported it to his party committee. After the party committee investigated, two specialists took the giant panda back to the mountains and released it deep in the forest.[72]

Mill workers were generally Han Chinese and the panda seems to have become ensnared in his trap accidentally. In this account, the government is portrayed as nobly delivering the giant panda to its home range. This story is listed between two other accounts of county successes, all of them cited to show that the behavior of people in the area had genuinely changed. Another account commended two commune members from a Baima commune, Muzuo, for their actions.

> In April of 1964, two members of the Muzuo commune team, Wang Chaolin and Ge Niu, encountered a giant panda and a takin while hunting. They both decided to follow the regulations and stop hunting and therefore received praise from higher-level authorities.[73]

How this story reached the authorities in a substantive enough form to elicit praise and be recorded is unknown, yet it showed that the provincial- and county-level authorities used praise as well as punishment to enforce the policy. A third account seems to have required an equal amount of faith on the part of the authorities.

> In February of 1965, a soldier from the Baima commune named Ge Ge went into the mountains to hunt. His hunting dogs surrounded a giant panda, but he did not shoot. Instead, he called his dogs back.[74]

All of these stories are about what did not happen, but from the perspective of the county government they were all success stories, since a giant panda was spared. Their main value as far as the county government was concerned was that they demonstrated to higher levels of government that the county was achieving residents' compliance to this new animal-protection law.

The reality of what such changes in policy actually meant to the people is gleaned through government documents in bits and pieces. We do not know the degree to which compliance was achieved, or the extent to which it was achieved out of fear of punishment, inspiration for reward, or a simple congruence of cultural beliefs with the new policies. Regardless of how enthusiastically or begrudgingly the local Baima people reacted to the new policies, the new policies clearly came with consequences and restrictions that the Baima had to confront.

The placement of the Wanglang reserve between the county border and the Baima village required that county officials regularly pass through the Baima community to run the reserve. This placement more comprehensively integrated the Baima community into the county and thus the nation. Zhong did not state that this was a specific goal; as an official entrenched in the county government and in the national project, however, he viewed such integration as potentially positive.[75] There was no consensus among the Baima about the reserve. For some, the presence of the reserve was undesirable. Others viewed it basically as insignificant. Some lamented that the reserve shrank hunting, herding, and herb-collecting areas.

Others expressed hope that the reserve's existence would inspire the building of a proper road to their village, indicating that those Baima people saw increased integration as more favorable than isolation. A road could facilitate much needed trade. In fact, as noted by a few Baima villagers, discontent with the reserve had more to do with how long it took to finally get a good road to the village rather than the impositions and restrictions the reserve created.[76] It must be noted, of course, that the road might simply have been seen as a safer complaint to make to an outsider who had stepped out of a county forestry bureau vehicle. At the very least, such sentiments indicated very mixed feelings regarding the reserve itself and a strong belief by some Baima that a good road would improve their lives.

The creation of the Wanglang panda protection reserve in 1965 demonstrated how a concept such as nature-protection policy could originate in the nation's top legislative body and be transformed into physical reality in the remote mountain ranges of southwestern China. The process by which the site was chosen highlights how provincial and local concerns influenced the location and design of the giant panda's protected space in ways that added value to the existing aims of the reserve. Examining the reserve's establishment also sheds light on standard procedural workings of the government of the People's Republic of China. Directives from the top and central offices did reach remote mountain hinterlands by way of a fairly direct relay of documents. The incorporation of provincial, local, and individual interests during the course of policy implementation both reinforces and belies the characterization of China during this era of history as operating under a dominant national government.[77] On the one hand, the PRC government assigned new value to wildlife and new restrictions and responsibilities to the local people. On the other hand, individuals and local concerns significantly influenced when and where these changes took place. Officials from China's top government body, ethnic minority hunters, Zhong Zhaomin, and others all participated in creating China's first giant panda protection reserve, each in tangible ways that demonstrate the importance of the responsiveness of individual and peripheral counties to China's broader national project.

A single individual, Zhong Zhaomin, was able to influence the implementation of national policy so that it could respond to local concerns. Of course, it helped tremendously that the needs of the county were not in conflict with the purpose of the nature reserve. Rather than offering another example of negotiation between the central and local governments, the story of the Wanglang reserve demonstrates that the central government gave a great deal of autonomy to the local government in implementing national policy. Furthermore, the nation and the locality both benefitted from this structure. Zhong's approach to establishing the reserve was simultaneously pragmatic and idealistic, and he found simple ways to address multiple concerns at once. He not only took his job of carrying out national policy at the local level seriously, but he also carried it out with a tremendous amount of nationalistic spirit. When asked how he could endure all the hardship that accompanied "eating in the wind and sleeping outside" to do his job well, he replied, "It was for the nation."[78] Moreover, the workers he recruited, such as Cai Li, who helped Zhong with early surveying work on the steep mountainsides, responded similarly when asked how they sustained themselves through the arduous physical labor.[79] With this enthusiasm, Zhong embraced a wide spectrum of associated tasks, including caring for pandas directly (see Figure 3.5). What is most striking is that Zhong was able to have such breadth and depth of vision to see the county, its forests, pandas, people, and the nation as part of the same organism, all ultimately sharing the same needs.

The central government continued to participate in this new space called the Wanglang Nature Reserve. Through this and other reserves, it made concerted efforts to better understand and protect the giant panda long before western scientists and international NGOs got involved. The national government allowed the province and county to take charge of the task of designating space and running the reserve, but it remained interested in and involved in the reserve's development. In addition to being a demarcated space, the reserve represents the bureaucratic process involved in creating a protected space, the significant labor invested in its creation, the deliberation over local concerns, and the interaction among the many people involved in the process. It also represents the activities and communications required in the perpetuation of the Wanglang reserve as an operating space. Of course the pandas also actively affected the placement of the reserve and continue to affect its management.

FIGURE 3.5 Zhong Zhaomin, founder of the Wanglang Reserve, is pictured here nursing a giant panda cub in the reserve office. Courtesy of Wanglang National Nature Reserve, Pingwu County, PRC.

The implementation of the national nature-protection policy also demonstrates the existence of a space in which the central government and the locality could maneuver. Room for maneuvering probably existed to a much greater degree in the remoteness of panda country than in more closely monitored spaces.[80] Nevertheless, the process of implementation shows how local actors could color and shape the policy that was enacted on the ground. In the end, the process of enacting nature-protection policy in Pingwu County took on the complexities of the landscape that it aimed to enclose, revealing a dynamic space that in turn served the protection needs of the pandas, the people of Pingwu, and the People's Republic of China.

4

Pandas Are Red

THE CULTURAL REVOLUTIONARY RISE OF THE PANDA AS BRAND AND SYMBOL

WITHIN A YEAR of the creation of the Wanglang Nature Reserve, China was enveloped in the chaos and political turmoil of the Cultural Revolution. Often described as the "decade of chaos," the Cultural Revolution period (1966–1976) is characterized by the dramatic upheaval of society in which Mao and his fellow radicals called upon the nation's youth, peasants, and workers to dismiss the authority of experts and take control of society. The resulting upheaval particularly impacted such sectors as science and medicine that largely based authority on expertise. Despite this context, a prestigious group of scientists and forestry officials gathered at the new reserve. Seventeen scientists had come all the way from Beijing to the mountains of northern Sichuan to represent the Institute of Zoology under the Chinese Academy of Sciences.[1] The Ministry of Agriculture and Forestry sent representatives from the national, provincial, and county offices to join the experts from Beijing. Once in Pingwu, the group also invited "eight hunters from the local minority group."[2]

A second survey of Wanglang was under way. This time, instead of seeking out a panda habitat, the surveyors were looking for pandas. This survey was in many ways an anointing of the reserve—affirming it as a functioning space that was engaged in real protection work. The most striking feature of the survey, however, was its timing, from 1967 to 1969, the height of the Cultural Revolution. While the rest of the country was awash in big character posters and revolutionary slogans, some members of the scientific elite were given the opportunity to spend time engaged in scientific research on the steep slopes in panda country.

This chapter focuses on the ways that the Cultural Revolution affected giant panda protection efforts and transformed the meaning of the image of the giant panda. Without undermining the complexity of the decade, it is also important not to ignore or dismiss efforts to engage in productive scientific work and the impact of these efforts on the development of nature protection in China. Discussions of protection history have portrayed the entire decade as a "black hole" for conservation, an era not only devoid of environmentally protective efforts, but smeared with environmentally destructive events.[3] There certainly was extensive destruction that occurred during these years that should not be downplayed. At the same time, many people also put forth a tremendous amount of effort to conduct research and pursue productive nature-protection work in spite of general discouragement of such efforts.

The first few years of the Cultural Revolution are commonly viewed as the most dramatic both in terms of zealous participation and devastating impact. Many intellectuals were sent to various places in the countryside to labor and be reeducated, including hydro-engineer Huang Wanli, who had opposed Mao's ambitions for damming the Yellow River, and biologist Pan Wenshi, who later became famous for his research on wild giant pandas.[4] The fact that this seminal panda survey team in 1967 included government officials and established scientists, had a clear research agenda, and openly reported the dates of its duration indicates that the government endorsed this project.

The image of the giant panda also grew in popularity during this time all over China. It became a noted and acceptable means of modernizing traditional art forms and glorifying the nation, growing from simply a precious species to becoming a "national treasure" and a point of pride for China and the Chinese people.[5] Both the scientific study of this animal and the growing ubiquity of its image in industry and art contributed substantially to this shift. The Cultural Revolution was not simply a period of significant popularization of the panda image; rather, the Cultural Revolution atmosphere launched the giant panda into the role of national icon.

Experts Escape to High Elevations

Wanglang Nature Reserve had only been operating for approximately two years when the survey team arrived in 1967 to study its esteemed residents. The reserve provided researchers with a mapped area of known topography where the existence of the panda was confirmed. The relatively close proximity of the Pingwu County Forestry Bureau and its officials' extensive investment in the reserve offered researchers a support network that included access to the land and local guides intimately familiar with the area. The main objectives of this survey were to understand the habitat, living habits, and distribution of giant pandas, as well as to gather any other fundamental information needed to improve the protection of these animals.[6]

The research team members took a comprehensive approach to achieving their objectives. They began their study with foot surveys of the reserve and interviews in the Baima villages just outside of the reserve border. Walking the landscape enabled them to become personally familiar with the habitat and identify basic signs of panda behavior, as well as evidence of any environmental degradation. Local inhabitants provided them with information on the area's flora and fauna, as well as observed behaviors and sightings of giant pandas. These interviews also provided researchers insights into preservation needs of the species.

The 1960s survey of Wanglang informed subsequent studies about seasonal activity and migration patterns, eating habits, and estimated population. Surveyors focused on the three major river canyons and three smaller canyons. In each area, researchers and guides spent seven to ten days surveying. The team also divided into groups of five or six people, each with a hunting dog, and surveyed different areas for twelve hours a day. Efforts to estimate the giant panda population were challenging because they did not have effective techniques to differentiate individual pandas. This sometimes meant that two pandas were counted as one and others were counted twice. These surveyors observed that, unlike typical bears, giant pandas do not hibernate. Through observation and interviews with experienced local hunters, researchers learned that pandas lived a solitary existence and that mothers raised their offspring alone. Surveyors also found fur in panda feces, belying the notion that pandas are completely vegetarian.[7] Observations of captive pandas confirmed the panda's penchant for meat. This specific finding influenced the decision to use meat to lure and bait pandas in later rescue efforts.

Although there was still a great deal to be learned about the giant panda after the 1967–1969 Wanglang panda survey, it was the first effort to study the wild panda in its natural habitat. A serious and influential scientific endeavor, it established much of the information now considered common knowledge about the giant panda and laid the groundwork for later studies.

Painting Pandas and Modernizing Tradition

Just as scientists discovered that the panda was a politically safe research subject during the Cultural Revolution, artists and manufacturers learned that rendering its image in art and as a logo was perfectly acceptable amidst the anti-traditional, pro-socialist movements that flowed from the fervor of the Cultural Revolution. By the mid-1960s, the giant panda became synonymous with modern China. Studying it, painting it, and mass producing it were all means of glorifying one of China's prized possessions and thus China itself. These actions affirmed China's ownership of this animal, and, through it, possession of another point of nationalistic pride.

Given how long the giant panda has roamed the region, it is quite surprising that it is virtually absent from traditional Chinese art forms that preceded the twentieth century. It was precisely because of this, however, that the giant panda was free of any associations with imperial China or "feudalistic" society that became distasteful during the Mao years and particularly problematic during the highly politicized era of the Cultural Revolution. As such, the giant panda became a modernizing force. In his famous "Yenan Forum on Literature and Art" in 1942, Mao called upon the masses to create new art to inspire and support the communist revolution. He stated,

> We should take over the rich legacy and the good traditions in literature and art that have been handed down from past ages in China and foreign countries, but the aim must still be to serve the masses of the people. Nor do we refuse to utilize the literary and artistic forms of the past, but in our hands these old forms, remoulded and infused with new content, also become something revolutionary in the service of the people.[8]

Even without being a symbol of revolution or socialism per se, the insertion of panda images into traditional art forms directly responded to the call for "new content" and thus served nationalistic and revolutionary artistic expression during the late 1950s and early 1960s. Artists incorporated the image of the giant panda most conspicuously through *guo hua*, or Chinese brush painting. Debates that emerged through and about art during the early 1960s "revealed a growing competition for authority within the cultural bureaucracy itself."[9] As a result, by 1963, "The intense frustration experienced by Mao, Jiang Qing, and other 'radicals' largely excluded from the post-Leap cultural order thus formed the context of a series of pre-Cultural Revolution purges."[10] These radical voices imposed a shift in the content of most art forms with the unleashing of the Cultural Revolution. The fact that giant panda images were a part of the artistic landscape during the years leading up to 1963 is significant because they survived these radical purges and became particularly popular in many genres immediately following them.

In 1961, the *Renmin ribao* (People's Daily) published a list of artists and their work that had "gained the Party's and nation's encouragement and the people's support." Among the art listed was a miniature sculpture of a giant panda fishing.[11] The same year, another image of the giant panda appeared in the paper, a new painting by famed artist Wu Zuoren (吴作人). Wu's painting was simply captioned "Panda, Chinese Painting" (熊猫, 中国画). The publication of another panda painting by Wu reinforced for the readers that the giant panda was a suitable form of new content in *guohua*, a genre of Chinese brush painting that dates back to the Song Dynasty (907–1279).[12] Because the giant panda, much like peasants, machinery, and members of ethnic minorities, is not traditional content for *guohua*, the paintings of pandas by Wu and other artists allowed this genre to be redefined. The endorsement of the panda as an artistic subject by the *Renmin ribao* (People's Daily) assured viewers that the panda was politically acceptable.

In 1963, the giant panda achieved even broader acceptance as an expression of Chinese national painting when the government issued national postage stamps featuring four brush paintings of giant pandas by Wu Zuoren.[13] The government issued these stamps, see Figure 4.1, one year after the State Council declared the panda a nationally protected animal. By this time Mao and his cohort had already begun asserting their authority over artistic content, so clearly they approved of it. Even as the art police became more radical and aggressive, they did not punish the panda for its increasingly high-profile appearances during the early 1960s.

During this same period the giant panda also became an international symbol of rare species. In 1961 the World Wildlife Fund adopted it as their emblem. The artist who created the WWF branding knew the animal was popular because pandas that were extracted from China by foreign trappers during the 1930s and 1940s had taken the world by storm. Even though the WWF had no panda program, nor any hope of starting one at the time, it felt the panda symbolized its work. Not inconsequently, during an era when color reproductions were very expensive, its natural black and

FIGURE 4.1 This 1963 government-issued stamp features brush painting by famous artist Wu Zuoren and reflects the government's participation in the growing affiliation between the giant panda image and the People's Republic of China.

white coloring and its simplicity as an image made it an ideal emblem for this young NGO focusing on wildlife protection.

In the context of China's politics during the years between the Great Leap Forward and the Cultural Revolution, the adoption of an image by a European-based international organization that sought both to benefit from and be a benefit for nature through philanthropic exchange could only have a damning effect on the image as a nationalistic expression. The characteristics of the WWF organization were antithetical to what came to be hallmarks of the Cultural Revolution. Yet even the WWF's adoption of the giant panda image produced no negative impact.

Deemed a politically safe image, the giant panda became an even more popular motif in various art forms and as a product brand during the Cultural Revolution. In this intense atmosphere, political attentiveness became critical to the survival of images, ideas, and those who introduced them, especially for those engaged in such a borderline bourgeois endeavor as producing art. Continued praise in newspapers of panda designs in many traditional art forms popularized their acceptance. This only added to the increasing ubiquity of its use by artists anxious to demonstrate their adherence to the political demands of the time and those who were seeking alternatives to some of the more standard artistic tropes of peasant and industrial images.

From handicrafts to high art forms, pandas became a means of modernizing tradition. They explicitly responded, according to period newspapers, to Mao's call to let "new things emerge from the old."[14] The increased association between pandas and the nation was repeatedly demonstrated by more artists creating new renditions of it and more companies incorporating it into logos.

Throughout the Cultural Revolution, China's main national newspaper, *Renmin ribao*, repeatedly celebrated decisions to incorporate the giant panda as a new design in traditional forms of arts and crafts. In Hubei, the new design of "panda eating bamboo" debuted in August of 1972 as a modern pattern in a form of textile handicraft. This panda design replaced such traditional images as flowers, birds, insects, and fish. The artist and designers who developed the design were also commended for producing a lifelike image.[15] This final note of praise was echoed in many articles about new art and panda designs, as it reflected a trend toward realism that was a highly valued component of socialist art. No longer were artists encouraged to copy the ancient masters, but instead were told to "sketch from life." Painting from nature could glorify the nation and foster the people's love for their country, but it was expected that the paintings would reflect what was actually seen.[16] This kind of art glorified China's vast nature in a way that was expressed in the grand scientific expeditions that *China Pictorial* magazine publicized for national and even international consumption.

Pandas appeared in many artistic genres in quick succession. In 1972, the renowned Jingdezhen porcelain factory in Jiangxi Province, which had become famous in 1320 for perfecting the blue pigment first used to decorate porcelain, launched new products that demonstrated that a traditional porcelain factory could overcome its own imperial heritage. The factory displayed its innovative techniques and technology by introducing two new, modern designs: an image of a panda and an image of a game of ping pong. Both designs were carved into porcelain by a skilled carver. *Renmin ribao* offered formulaic praise for the images' realism.[17] The popularity of these designs, attributed in part to their realism, was important for demonstrating that these new designs succeeded in fulfilling the purpose of new art "for the people."[18]

The next year an article titled "Updating traditional arts and crafts" featured the introduction of new silk brocade patterns in a Guangxi fabric factory. As another explicit response to Mao's call for the "new to emerge from the old," the brocade factory created a "panda with emerald green bamboo" design for quilts.[19] Again, praise emphasized the realism of the pattern and its popularity with "the masses" to demonstrate the success of this particular factory in breathing new life into traditional handicrafts. In yet another case, a bamboo carver who already had achieved recognition in 1972 by spending time in the countryside and producing realistic designs that evoked "peasant life and struggles" was praised for his new bamboo carvings of pandas.[20] This designer went to the Beijing zoo to observe the giant pandas' movement and behavior in order to carve many lifelike poses of pandas on fans, tea leaf boxes, cigarette boxes, and other daily-use items. The carver "improved upon" the ancient art form with his meticulously carved lifelike images of the giant panda, which were "well received by the masses."[21]

This repeated endorsement of the giant panda indicated that it was important to educate the masses in order to buttress the acceptability of the panda as a modernizing and revolutionary motif. In addition to being suitable "new content" for socialist art, recurring testaments reinforced that the giant panda was also popular and therefore suitably socialist.

Manufacturing a Symbol

Just as the giant panda infused old art forms, it also was increasingly used as a means of advertising China's self-proclaimed advancements in science, technology, and industry during the Cultural Revolution. Industry first seized upon the panda image during the new socialist era. In 1956, an ambitious Chinese electronics company aimed to advance electronic radio technology within the domestic and international markets. The company sought a name that would evoke the Chinese nation. After rejecting numerous other suggestions, company representatives unanimously agreed upon "熊猫" (*xiongmao*), or "Panda," as its brand name. A company representative remarked on this choice in a newspaper article celebrating the company's fifth year of successful production of radios and other electronics: "The 'panda' is China's most famous precious animal—upon seeing 'Panda' one would, therefore, immediately know that it is a Chinese product."[22]

The panda subsequently became an increasingly popular brand name for a wide variety of products. During the early 1960s, an Inner Mongolian dairy promoted its "Panda" brand of condensed milk and butter.[23] The name did not reflect any characteristic of the product or the region in which it was produced, since pandas are not native to the region of Inner Mongolia. This choice was an attempt to cater to the broader Chinese market and tie Inner Mongolia to the nation of China.

In 1962, the panda modeled advancements in plastics with the debut of a supple, plastic toy panda. The company promoted plastic as a good material for toys: "If it is dropped or falls it doesn't break." It portrayed the panda as an ideal form for a toy, "fat and wide with big, round, black, eyes . . . it lets out a 'jihu jihu' sound that makes children laugh loudly."[24] Nothing about the panda or its image implies anything especially linked with advancements in plastic technology or other modern industries. Associating milk and toys with the giant panda, however, reflected wholesome associations between the giant panda and the main consumers of these products—children. The PRC government itself also turned to the panda as a means of branding technology that China wanted to promote as exemplifying national industry. The animal was chosen to represent the laser technology developed during the Cultural Revolution. When New Zealand Prime Minister Sir Robert Muldoon and his wife visited China in 1976, they took an interest in learning more about the mathematical program that students, workers, and instructors at Qinghua University had developed for the laser housed on campus. Their Qinghua guide was anxious to demonstrate the quality of this technology, which "combined the efforts and talents found in the combination of the classes and therefore illustrated the superiority of this cooperative, class-conscious approach."[25] The Chinese guide used the laser to etch the image of a panda on a piece of glass, which he then presented to the New Zealanders as a memento of China and the scientific advancements achieved through "Cultural Revolutionary spirit and hard work."[26]

In spite of the apparent pride invested in new products, criticism that the Cultural Revolution had wrecked China's industry, stifled its technological advancement, and halted development in science was growing. The counter Cultural Revolutionary slogan "the present is not as good as the past" [*Jin buru xi* 今不如昔], famously attributed to Deng Xiaoping, challenged the concept that politics were paramount to expertise. One news article rebutted this criticism by showcasing the exemplary company, Panda Camera. This article insisted that China did progress technologically during the Cultural Revolution by highlighting the strong rate of production and diversity of new cameras manufactured. The author argued that in 1976 camera production rates were twelve times those of

1965 and the number of camera models had increased four-fold since 1965.[27] He continued with the history of this particularly revolutionary camera company that "persisted in class struggle and diligently carried out 'The Charter of the Anshang Iron and Steel Co.' as its guiding principle of continually pursuing the 'In Industry Study Daqing' movement."[28] The article characterized this camera company as continually engaging in criticism of such flawed attitudes as "wait, depend, need" and other such examples of "cowardly ways of thinking." Instead, the young camera assembly workers, leading cadres, and technicians, who previously had no camera production experience, accepted the difficult task handed to them to produce a trial camera model.

> Putting proletarian politics in command, they carried out self reliance and kept up the practice of plain living and hard struggle with Revolutionary spirit, simultaneously rebuilding the factory and doing the trial productions. Working and studying, they quickly grasped the art of camera manufacturing and production. In just over three years, they successfully produced "Panda" brand cameras, which have undergone national-level testing and are already in mass production.[29]

The strong political rhetoric of Revolutionary spirit overcoming the need for trained expertise reflects the atmosphere of this highly politicized decade. In this rendition of events, the successful "Panda" camera company explicitly represented the political values of the Cultural Revolution.

The giant panda was also the subject of China's first full-length, color science education documentary, entitled simply *Panda*. This project combined nature, art, and technology and showcased China's advances in film production technology nationally and internationally.[30] Set in the Wanglang Nature Reserve, the film simultaneously offered a display of China's biological uniqueness and natural beauty.

Articles reporting on the success of *Panda* and other films produced during the Cultural Revolution period raved about the latest developments in color film processing, lifelike color, and cinematography.[31] Mao boasted that *Panda* and the other contemporary Cultural Revolutionary film entries "can satisfy most needs [nationally], but many can also be exported." Subsequently reports affirmed that these films had been well received at international film festivals in India and Italy.[32] *Panda* was the only film among the group of the eight submissions that China sent to both festivals. The other Chinese films featured in these festivals specifically reflected China or Chinese culture in some way, such as *Beijing Amateur Martial Arts Troupe* and the famous Cultural Revolution opera *The White-Haired Girl*. The documentary on giant pandas, unlike films on the period's modern opera and revolutionary ballets, was not an explicitly revolutionary topic. The subject matter of the film, China's prized precious species, however, was clearly a nation-glorifying choice. The Chinese government also offered *Panda* with some of these other films as state gifts to Iran, Iraq, and other foreign countries.[33]

The success of *Panda* was hard-won by its filmmakers. The crew began working on the project in February of 1973. They trekked through the steep and dense bamboo-covered Min Mountains of the Wanglang reserve for nine months before finally seeing a giant panda. By this point they had determined that the whole project would be a failure if they had to rely solely on the elusive mountain creature to fill the frames of their film. With government approval, they caught and tamed a young wild giant panda "actor," which they named Wei Wei. In the process, a Baima Commune party secretary sustained serious injuries trying to protect the film crew by fending off the young panda's mother in what turned into a human-panda brawl.[34] As it appeared on screens around the country toward the end of the Cultural Revolution, the film gave its viewers a peek into the Wanglang Nature Reserve. Through this film, the public gained rare access to the panda's beautiful and rugged environment. Since the film was shot and produced with domestically manufactured technology, it also highlighted the success of technology produced during the Cultural Revolution for the international community.

In 1973 China finally published information from the 1967 giant panda survey in Wanglang in the popular *China Pictorial* magazine. Opening with two huge photos of Wanglang's wild giant pandas, *China Pictorial* allowed readers to gaze at the photographs of China's precious panda in the wild. The article offered a short summary of the reserve's wealth in wildlife, while saving the in-depth discussion of survey methodology and results for publication in an academic journal.[35] The article glorified China's expansive and diverse topography, as well as the rigor and hardiness of the Chinese scientists involved in the reserve's fieldwork. The glorification of field science was a magnification of a long-standing heroism associated with discovery and fieldwork and, in the context of the Cultural Revolution, a vilification of the lab and its experts.[36,37] Even the academic journal article detailing the scientific methodology and findings echoed this sentiment. Although an extensive normalization process was under way during these later years, the atmosphere in China remained politically charged. Blatant revolutionary rhetoric continued to fill the preambles of a wide spectrum of academic studies even beyond the death of Mao.[38]

The Cultural Revolution decade put incredible strains on the Chinese people as they tried to navigate the shifting politics and safely express nationalism. The fact that the giant panda emerged as an expression of nationalism with such unproblematic consistency demonstrates how well suited it was to the challenging task of representing unwavering Chinese nationalism during such a volatile time. Probably its greatest asset during the Cultural Revolution was that it could not be linked to either the exploitative nature of China's dynastic history in its time of strength nor to the embarrassment of China's weakness during the long decline of the last dynasty. It could offer the nation an image that was simultaneously distinctly Chinese and distinctly modern—and could stamp these qualities on everything associated with its image. Eminently malleable, the image of the giant panda was able to survive the many political shifts that deemed other images unacceptable. Not even the image of Mao Zedong himself was able to serve as a national icon with as much consistent support and enthusiasm as the giant panda. Not only was the giant panda untainted by political associations, but it also survived the Cultural Revolution unscathed. The Cultural Revolution enabled the giant panda to transform from an expression of nationalism into an acknowledged symbol of the nation. Yet when people reflect on the Cultural Revolution and the trauma associated with it, they do not think of the giant panda or its image. Although the Cultural Revolution was a central force in popularizing the image of the giant panda, the giant panda remarkably escaped any negative associations with the Cultural Revolution itself. The political power of the giant panda was its innate ability to exude an apolitical image.

5

Panda Diplomacy

ANIMAL AMBASSADORS AND THE WILD POPULATION

Ambassadors and other dignitaries were on hand Friday to celebrate the 10th anniversary of a popular Washington couple who have entertained millions during their decade here. The scene was the National Zoo and the party honored a pair of pandas, Hsing Hsing and Ling Ling, a gift to the nation from the People's Republic of China.

—ASSOCIATED PRESS, April 16, 1982

FROM THE TIME Hsing Hsing and Ling Ling arrived in the United States in 1972, they were not only honored by the presence of ambassadors and other officials, but were also likened to ambassadors themselves. Subsequently, pandas have been honored as high-ranking celebrity ambassadors in every country that China has chosen to grace with a gift or loan of pandas.

As the Cultural Revolution drew to a close, pandas were bestowed upon foreign nations in a grand and greatly appreciated act of international friendship that came to be known as panda diplomacy. When most other aspects of the People's Republic were threatening or, at best, beguiling to many foreign countries, live giant pandas served the Chinese government as invaluable tools for putting a friendly face on China. The giant panda, once a symbol of extreme nationalism and self-reliance, became useful to China by helping the nation reach out to engage the world.

China did not have state-appointed human ambassadors in many of the countries to which they offered the animals because the majority of countries did not have established diplomatic relations with the PRC. Even after its admission to the United Nations in 1971, many countries officially continued to recognize the Republic of China in Taiwan as the only official government of China. The arrival of the two pandas built on the important foundational steps that Nixon and Mao took to warm relations between the two countries during their seminal meeting that same year. The gift, following President Nixon's visit to China, began a new era in China's foreign relations that has been punctuated dramatically with well-publicized panda diplomacy.

The term *panda diplomacy* emerged during the early 1970s when the Chinese government bestowed pairs of giant pandas as offerings of "goodwill and friendship to the people" of the recipient country.[1] Some have seen the giant pandas used more as tools, either to motivate or reward countries for engaging in China-friendly relations. This perspective is most famously represented by the offering of pandas to the United States and Japan in 1972 as each former enemy state began new types of engagements with China. When pandas were offered to Taiwan in 2005, objections rose that it was a "charm offensive" to manipulate the people of Taiwan.[2] Long after China ceased giving pandas as gifts in most parts of the world, panda diplomacy was used to describe not only the various panda loan arrangements that followed, but also the apparent political maneuvering and gratitude that seemed to surround such loans. Even when there seemed to be no apparent relationship between the pandas and politics, they became entwined in the public mind. For instance, in 2010, articles in the international press speculated that the decision to send the first successfully captive-bred giant panda in the US National Zoo "back" to China was related to tensions between the two governments surrounding the visit of the Dalai Lama to Washington.[3] In fact, a component of all twenty-first century loans of pandas is that any offspring of the loaned pandas belong to China and will be sent to China when the cub is an appropriate age. The decision to send the panda cub to China was unrelated to the Dalai Lama's visit. The term *panda diplomacy* can also refer to image-softening and affection-generating effects among the panda-viewing public in the foreign nations to which they are sent. In the same way that other types of cultural exchanges are meant to generate understanding through exposure and culturally focused education, panda exhibits offer foreigners an opportunity to appreciate something that is intrinsically Chinese, yet is not associated with any perceived threatening aspect of China's political or economic culture.

Much of the giant panda's potency as a diplomatic tool is directly related to its appealing appearance. The aesthetic enjoyment experienced when looking at the animal is enhanced by the sense that the opportunity to see one in person is rare and special because it exists naturally only within the territorial borders of China and is famously rare. Of course, by the time that giant pandas began traversing the globe as animal ambassadors, they already had been popularized internationally by the World Wildlife Fund. It certainly is possible that the panda's use as a symbol of the WWF for the decade prior to China's early offerings of the animal as a state gift helped magnify the effect of the gifts by making the image familiar among certain circles worldwide. This is compounded by the fact that during the early 1970s it was incredibly difficult to gain any access to China, let alone the remote region where wild pandas lived. The WWF's use of the giant panda as their symbol certainly did not hurt the PRC's efforts to use the animal as a diplomatic tool in subsequent decades.

Panda diplomacy has had a profound effect on the wild population and China's domestic conservation policy. Concerns about the well-being of wild pandas have, in turn, affected the terms of diplomatic panda exchanges and fundamentally contributed to debates surrounding specific gifts and loans. Although the international community has been instrumental in panda preservation efforts in recent decades, China's domestic efforts predate any contribution by outside organizations, individuals, or nations to the well-being of this unusual species. Examining the history of panda diplomacy traces the transformation of China from a country that desperately solicited international friendship to one that has become confident in its ability to exert influence on a global scale.

The Beginning of Panda Diplomacy

The 1972 gift of the pandas to the United States was perhaps the most famous example of panda diplomacy, but was by no means the first. Many Chinese and English language articles, scientific and popular books, and news reports date panda diplomacy as far back as the Tang Dynasty (618–907).[4] According to these many accounts, a Tang-Era emperor of China, possibly the famous sole female monarch Wu Zetian, sent two live pandas and seventy panda pelts to the emperor of Japan. This story elicits a tremendous amount of interest because it indicates that the panda was seen as an object of value more than a thousand years ago. The Japanese text on which this story is based, however, not only does not describe pandas, but also does not describe an exchange between the Tang emperor of China and the Japanese emperor. Rather, it states that a Japanese lord went on an expedition to the northern territories and returned with two live brown bears (likely *Ursus Arctos*) and seventy brown bear pelts.[5] Unfortunately the Tang-Era gift of pandas to Japan turns out to be a case of unintended myth-making through a mistranslated and misinterpreted text. The widespread reporting of this event, however, demonstrates that people, both within and outside China, desire to add legitimacy and mystique to current practices with this historical precedent.

The first act of panda diplomacy was, perhaps ironically, an expression of appreciation to the United States for its assistance in China's war efforts against Japan. In 1941, Madame Chiang Kai-shek and her sister Madame H. H. Kung presented two giant pandas, later named Pan Dee and Pan Dah, to the United States in a gesture of gratitude.[6] Even while the war was still under way, these pandas made a harrowing journey through enemy territory and across the Pacific. At the time, not much was known about giant pandas in China or abroad. Madame Chiang Kai-shek and the Kuomintang (KMT) government knew that the American people would like them due to the spectacular popularity of the 1930s and 1940s pandas that Harkness and Tangier-Smith brought to the West.[7] By the time the 1941 gift pandas arrived in the United States, only four of the original twelve giant pandas on exhibit overseas were still alive, two in Chicago, one in St. Louis, and one in London. The female (Pan Dee) only lived for four years, but the male lived for ten years at the Bronx Zoo.[8]

After the Chinese Communist Party took over the governing of mainland China, it initially followed suit in rewarding political friendships with this spectacular token of gratitude. The PRC offered its first gift pandas to the fellow socialist nations of the Soviet Union and North Korea. It gave two pandas to the USSR, one in 1957 and one in 1959. It then gave two to North Korea in 1965, and two more in 1971, apparently as replacements.[9] The pandas given to the Soviet Union were not celebrated; only a single, short article appeared in the *Renmin ribao (People's Daily)*. The giant pandas bestowed on North Korea also attracted little attention. They were sent with a group of other animals as part of an "ongoing exchange of animals and plants to contribute to the ever closer friendship between the peoples of China and North Korea."[10] Along with two pandas, the mayor of Beijing gave a hippopotamus, a mountain goat, and a white peacock to the city of Pyongyang. That they were sent as part of municipal exchange from the mayor indicates that at that time they technically were not even viewed as state gifts. The *People's Daily* does not so much as mention the 1971 replacement gift of pandas to North Korea, which were transferred to Pyongyang from Beijing the same day that one of the earlier pandas died.[11] Such gestures of friendship between these neighboring socialist countries that had united in war against the United States during the previous decade were not surprising at that time, but they set a precedent for more newsworthy gifts to follow.

In spite of the popularity of zoo exhibits of pandas in the United States during the 1930s and 1940s, American zoos had little hope of replacing these desirable animals during the 1950s. Senator McCarthy's "Red Scare" politics openly politicized pandas by including them in the embargo on trade with China. After enumerating the many thwarted efforts by China to establish cultural exchanges with the United States, Chinese Representative Zhang Xiruo poked fun at the US policy that banned importing giant pandas from China in a speech on foreign relations.

> Americans would really love to obtain our native giant pandas, but their government uses the excuse that no exchanges with "Red China" are allowed and therefore, prohibits the import of giant pandas. American imperialists are frightened of even our giant pandas. Does this not expose their [the US government's] terrified cowardliness?[12]

Years later, *New York Times* writer Joseph A. Davis commented on England's 1958 acquisition of a giant panda from China: "Chi Chi the giant panda might have become an American citizen had not the Government banned her entry in the late 1950s as 'a product of Communist China.' Instead she flourished—and continues to flourish—as the pride of the London Zoo."[13]

Prior to offering giant pandas as a state gift to the United States in the 1970s, the Chinese government hosted President and Mrs. Nixon on a historic visit. Mrs. Nixon went to see pandas in the Beijing Zoo between a visit to the famous Summer Palace and a trip to the kitchen of a Chinese restaurant.[14] The giant pandas were presented as a unique natural phenomenon housed in the only environment where they had successfully reproduced in captivity under the care of Chinese scientists. The government offered a pair of pandas to the United States before the Nixons left China.[15] Less than a month later, two American musk oxen arrived in Beijing as a reciprocal gift.[16]

Public allusions to the animals serving as diplomats filled reports on the events in both nations. The oxen were greeted as if they were heads of government—with high-level state representatives receiving them. Beijing's Revolutionary Committee head, Wu De, met the musk oxen on arrival and "escorted" them to the Beijing Zoo. Head secretary of the

Beijing Municipal Revolutionary Committee, Huang Zuozhen, was also present. The director of Washington's National Zoo and US Congress Representatives accompanied the oxen to China. The Chinese press described the musk oxen as a gift from United States President Nixon to the Chinese people.[17]

The giant pandas arrived in the United States with even more pomp and official celebration than the musk oxen had in China. Transported on a US Air Force plane, they landed in Washington, DC under "the kind of security precautions tendered to visiting heads of state."[18] Another news article characterized the preparation for the arrival of the pandas as so full of import that it was "as if it were Mao Tse-tung himself."[19] A welcoming ceremony was held for the giant pandas at the National Zoo attended by the First Lady, numerous government officials, and more than 1,000 dignitaries.[20] (See Figure 5.1 for an image of Hsing Hsing and Ling Ling in the National Zoo.) Once they were introduced to the general public, the number of visitors to the panda exhibit reached 1,000 an hour.[21] This

FIGURE 5.1 Hsing Hsing and Ling Ling in their enclosure at the National Zoo in Washington, DC. Courtesy of Smithsonian Institution Archives. Image # 96-1378: Photograph by Jesse Cohen of giant pandas Hsing Hsing and Ling Ling at National Zoo in 1985. BW image.

outpouring of interest was consistent with a growing curiosity about China over the previous decades. Press in both countries offered the Chinese and the American people a friendly image of the other nation. A *New York Times* writer noted: "Until the two countries exchange ambassadors . . . it is not unreasonable that one pair of American musk oxen and one pair of Chinese pandas be spared for accreditation to each other's capitals as temporary chargés d'affaires."[22]

The exchange itself, however, had not completely displaced old tensions between the two countries. Evidence of lingering suspicions surfaced about the exchange. Some Americans feared that China was going to trick them and send two male or two female pandas rather than a pair that could mate.[23] Another rumor developed that the female was going to be well past her reproductive prime. Nonetheless, on April 16, 1972, two eighteen-month-old giant pandas arrived, one male and one female.[24] That suspicion of the PRC and their intentions behind the gift lingered in the United States even after the arrival of Hsing Hsing and Ling Ling is evidenced by the cover of the *National Lampoon* humor magazine (see Figure 5.2), which featured a giant "Trojan" panda with China's armed People's Liberation Army soldiers running out of a trap door in the bottom of it.[25]

The Chinese actually had much more to complain about. Although the musk oxen Matilda and Milton were charming in their own way, they were not even a semblance of a fair exchange for pandas. Bison would have had more symbolic virtue. Even US National Zoo director, Theodore Reed, admitted that they were no comparison to the giant pandas: "Reed was practically hugging himself with delight as he described the two [pandas]. 'You're going to love these animals,' he predicted. 'Frankly, I just don't think musk oxen have the sex appeal pandas do. You like musk oxen, but pandas can steal your heart away.'"[26] Moreover, the musk oxen suffered from some sort of skin problem and were not placed on display for several months. Song Qingling, Sun Yat-sen's widow, aired her view of the exchange frankly: "'We got a bad deal.' She said with a laugh, 'That's Nixon for you.'"[27] Fortunately, about ten days later the musk oxen were able to meet enthusiastic crowds in Beijing.[28]

Subsequently China made panda gifts to other countries as part of Mao Zedong's conscious repositioning of China's place in the world. The social, political, and economic state of China in the 1970s made its need to create and foster friendly relations overseas increasingly dire. Though formerly disinterested in participating in international governmental organizations

FIGURE 5.2 Reflecting the truth in all humor, this cartoon depicts the genuine trepidation among many US citizens about China's state-gift pandas in 1972. The concept of the Trojan panda resurfaced decades later when China offered a pair of pandas to Taiwan as a gift in 2005. Cover art created by Michael Gross, *National Lampoon* magazine, July 1972.

from 1971 to 1976, China's "IGO [international government organizations] membership expanded from 1 to 21," showing an intensive shift in policy.[29] These panda gifts were yet another indicator that China was embracing a new relationship with the international community.

Patching the Rifts between China and Japan and Beyond

If ink spilled in the press was any indication of the respective importance placed on China's newly developing international relationships, the initiation of normalizing relations with Japan was clearly the most significant of the era. China's *People's Daily* newspaper ran three times the number of articles discussing the giant panda in the context of the new relationship between Japan and China than it did concerning that between the United States and China.[30] Sino-Japanese historical conflicts were both more direct and more deeply rooted than Sino-American ones. China and Japan had recently fought a brutal war on Chinese soil, with such wartime atrocities as the Rape of Nanjing and the human experimentation under Unit 731 still vivid in the memories of Chinese survivors. As China and Japan had deeper wounds, the expressions of a new era of political friendship between the neighboring countries and the value placed on the gift of the pandas were comparably higher.

In September of 1972, newly elected Premier Tanaka Kakuei visited China with the intention of establishing dialogue between the two countries. After the signing ceremony of the Sino-Japanese joint communiqué, Zhou Enlai expressed China's desire to give the Japanese people a pair of giant pandas.[31] When the two pandas arrived in Tokyo, security was high and riot police guarded them from hundreds of curious spectators.[32] The list of Japanese government delegates who attended the welcome ceremony was long, including the commanding officer, head secretary, Japan's minister of foreign affairs, and representatives from several political parties. The Chinese delegation to the welcoming ceremony included, among others, the Beijing public works representative and China's foreign trade representative. The Japanese-Chinese Friendship Association, the Japanese-Chinese Cultural Exchange Association, the Japanese International Trade Association, and the Japanese-Chinese Fisheries Association also were represented at the ceremony.[33] The state-gift pandas clearly exemplified a dramatic shift in political, economic, and cultural relations between two former enemy states.

The trade interests represented at this welcoming delegation are striking. In fact, Japan and China both had strong vested interests in expanding trade. The two countries had initiated stilted efforts to trade under the difficult political circumstances of US occupation of Japan during the 1950s and pursued an expansion of these efforts during the 1960s.[34] By 1965,

Japan had surpassed the Soviet Union as China's leading trading partner.[35] There was significant interest among Japan's population and politicians in further expanding these trade gains. Cold War politics and the tensions between the United States and China, however, had hampered Japan's ability to pursue more open relations with China publicly. Many in Japan saw this as persistent submission to United States hegemony well after occupation had ended.[36] Nixon's visit to China blindsided Japan's government.[37] Once Japan recovered from the "Nixon Shock," as this event came to be called in Japan, many were relieved because it opened the door for Japan and China to more aggressively pursue their own normalization process and expansion of trade.[38]

The Tokyo welcoming ceremony for China's pandas set a tone of effusive friendship. After thanking the Japanese people for the mountain cherry and larch saplings that previously had been sent to China, China's head of public works, Zhang Shaoguang, said, "The establishment of relations between China and Japan begins a new chapter in the interactions between the two countries. This mutual giving of gifts writes a beautiful page in this new chapter."[39] The day that the panda exhibit opened to Tokyo visitors, 18,000 people were lucky enough to briefly file past the panda cages. More than 100,000 had to be turned away.[40] *Chūgoku Fībā* (China fever) and *Panda Fībā* (panda fever) quickly spread in Japan. Japan's politicians were quick to take advantage of it. Prime Minister Tanaka, graced with the good fortune of having his name attached to the Sino-Japanese joint communiqué, adopted the giant panda as the LDP (Liberal Democratic Party) mascot, using campaign slogans such as "Don't miss the China boat" to promote his party in Diet elections that year.[41]

One year later, a Japanese-Chinese friendship summer camp was set up for young people. Japanese college students were reported to be writing Japanese-Chinese friendship plays performed in Chinese and singing Chinese revolutionary songs. Japanese children participated in a Japanese-Chinese friendship program called "Pandas come to our home in the hidden plateau."[42] A "Cherish Pandas Committee" was established by Japanese people enthusiastic about the friendship between Japan and China to commemorate the one-year anniversary of the normalization of relations between the two countries. The group invited the Chinese-Japanese friendship association head, Liao Chengzhi, and the vice chair of Chinese International Trade Committee, Liu Xiwen, to Japan to speak in November of 1973, after which they were to go to Beijing and express further gratitude for the panda gift.

Seeing just how resoundingly popular the gift of giant pandas was in the United States and Japan, the Chinese government decided to continue this unique strategy of diplomacy. Recipient countries continued to embrace these animals with open arms and to honor them with impressive receptions. Diplomatic formality prevailed wherever state-gift pandas landed. In France, the Chinese ambassador, head of the Ministry of Education in France, a representative of France's president, the prime minister's wife, the head of the natural history museum in Paris, and the head of the Paris Zoo all attended the welcome ceremony for Paris's new pandas.[43] Unlike the places where the giant pandas were helping the two countries overcome troubled relations, in France the pandas represented the "traditional friendship between the French and the Chinese people."[44] England honored the giant pandas at the London panda ceremony in November 1974 with the presence of the newly formalized Chinese ambassador to England, Song Zhiguang, the Queen's husband, Prince Philip, and the former prime minister.[45] In 1975, Mexico's minister of foreign affairs, vice head of Finance and Public Credit, the Mexican-Chinese Friendship Committee Chair, and many other officials did not wait for a formal zoo ceremony but actually went to the airport to meet the pandas.[46] The presence of state officials to greet the pandas demonstrated a strong recognition of the gravity of the gift from China and an acknowledgment of its significance as a symbolic solidification of interstate relations. All told, nine countries received twenty-three state-gift pandas from the founding of the PRC through the end of 1982.

Other cultural exchanges that followed were then associated with the panda. Ping-Pong diplomacy had actually preceded panda diplomacy, but subsequently the two became linked. In 1971, China invited the United States Ping-Pong team to China. When the Chinese Ping-Pong team accepted a reciprocal invitation to the United States, they arrived in close proximity to the giant pandas. The Chinese table tennis team's visit included a trip to a New York City elementary school, which was decorated with colored pictures of giant pandas under sign that read, "Thank you Chinese friends."[47] When the Chinese youth Ping-Pong team went to Japan in late 1972, special mention was made that the same official who accompanied the state-gift pandas to Japan, Zhang Shaoguang, also accompanied the Chinese team.[48] Both the panda and Ping-Pong provided friendly and non-threatening ways for China to initiate new interactions with foreign powers, including former enemy states, and to have these overtures reciprocated.

Representations of the giant panda returned to China in many forms. Recipients of state-gift pandas offered representatives from China a variety of renditions of pandas—photos, drawings, paintings, toy pandas—to express their appreciation for the animals and the diplomatic friendship that they represented. Panda images continued to be exchanged as reminders of the grander gift. Just as schoolchildren in the United States had displayed their colored drawings of pandas for the Chinese table tennis team, Japanese schoolchildren offered panda paintings to Liao Chengzhi, the Chinese head of the Chinese-Japanese Friendship Committee, and other Chinese guests invited to Japan's 1973 cherry blossom festival.[49] When Liao Chengzhi met with Prime Minister Tanaka five days later, he received three glossy color photographs of Lan Lan and Kang Kang, the pandas who had arrived the previous year.[50] In June, a Japanese group replaced a traditional temple-shaped float that was the centerpiece of an annual parade with a panda float for the benefit of Chinese observers.[51]

The giant pandas' ability to charm the citizens of their recipient countries also had positive influence at home. The publicity, enthusiasm, and interest that these panda gifts inspired among the general public of host nations strengthened the appreciation for the animal among the Chinese people. At the time of President Richard Nixon's visit, wild-animal-protection officials and members of the Ministry of Forestry coined the term "national treasure" (*guobao* or 国宝) while brainstorming a way to express the significance of this gift China was offering to the United States.[52] This formalization of the giant panda's status as official symbol of the People's Republic of China and an icon of international friendship with China inspired increased interest in and attention to the wild-panda population, with both positive and negative results.

The Wild Side of Diplomatic Work

The use of pandas as state gifts to foreign countries led to an increase in the extraction of pandas from the wild, which in turn accentuated the necessity for their protection. Pandas also became more prized and desired at home, leading to a sudden clamoring among domestic zoos to obtain pandas of their own. China's forestry officials and protection advocates struggled to respond with a policy that could offer a balance between accommodating the panda's new political significance and limiting the strain on the species.

The first successful domestic giant panda exhibits were the three giant pandas that arrived at the Beijing Zoo in 1955. They were overwhelmingly popular with the Chinese public and proved to be robust enough to survive in captivity. Between 1953 and 1971, seventy-seven giant pandas were captured for placement in zoos across the country. Of these, thirty-six survived and were exhibited in eighteen cities in China.[53] By the time of Nixon's visit, the ranks of thirty-six captive pandas were strengthened slightly by captive-born pandas. Twenty giant panda births had taken place in captivity, a remarkable achievement; unfortunately, only three survived. This meant that China had thirty-nine captive pandas from which to draw for its new program of state gifts.[54] This would seem sufficient for China's initial offerings, but domestic zoos seemed disinclined to give up their popular exhibits. Only two of the pandas already in captivity in 1972 later became state gifts to foreign countries (one to the United States and one to Japan). China's main panda resource, therefore, was the wild. With the exception of one captive-born panda given to Spain in 1978, the eleven subsequent state-gift pandas and three replacement pandas (for Japan and North Korea) all were captured during the remaining years of the 1970s.[55]

Although there had not yet been extensive national surveys of wild pandas to assess how many existed in the mountains of China, it was broadly accepted that the animal was quite rare. The fact that the Chinese government had placed giant pandas under national protection, along with eight other animals, as early as 1959 indicated that it was concerned about their well-being and perpetual existence.[56] Little if anything was actually done to protect them, however. When the government expanded its list to nineteen animals in 1962, it showed a renewed interest in wildlife protection. In spite of a lack of scientific data, subsequent information indicated that the government had been correct that the giant panda was rare. Because this information was based on discussions with people who lived in the vicinity of the animal, notations in forest surveys, and actual efforts to catch the animal, there was no way to assess the ability of the wild panda population to withstand the level of extraction to which it was being subjected. The broad recognition of the giant panda's precarious existence was also evident in the repeated pledges by recipient countries to take good care of the giant pandas and the large budgets assigned for this task.[57] Replacements were not expected if the pandas did not survive in the foreign zoos, although some were given. The offering of pandas to other nations, explicitly as "protected" or "precious and rare" animals, and as a "national treasure," also drew new interest to the protection of the animal itself and its status in the wild.

The number of giant pandas captured during the remaining years that the PRC maintained the practice of giving state-gift pandas to foreign countries in fact far exceeded the number of panda gifts sent overseas. The vast majority of newly captured pandas were thus sent to domestic zoos.[58] Chinese forestry officials feared that trying to meet the upsurge in demand for pandas from domestic zoos after sending pandas to the United States and Japan would devastate the wild population. As illustrated in Figure 5.3, between the years 1972 and 1980, ninety-two pandas were captured, with the peak year being 1972. Seventy-six of these captured pandas went to twenty-one domestic cities and towns, although not all captured pandas survived long enough for exhibition. During the 1970s, thirty cities and towns in China put pandas on display.[59]

Many scientists and forestry officials involved in wild animal protection were concerned about the increase in wild panda captures. Forestry officials began to fight this trend of extraction around 1973, a time when the radical revolutionary government was encouraging its citizens to return to work during the final years of the Cultural Revolution. Even with this move toward normalcy, scientists still had to remain politically cautious in framing their projects. They were further challenged by the fact that China lacked the resources necessary to assess whether or not its panda population could withstand rising extraction demands.

In 1973 the Ministry of Agriculture and Forestry showed a renewed attention to precious species protection and the rational use of natural resources. This is evident in a sudden proliferation of policy documents, ministry meetings, and drafts of suggestions for policy implementation. National, provincial, and county offices of the Ministry of Agriculture and Forestry injected new energy into their stratified approach to nature protection. That November, the Sichuan provincial office organized a special province-wide meeting to address the issues of precious animal resource protection and surveys. A Provincial People's Government official, Han Zhengfu, described an excessive demand for pandas in domestic zoos to an audience that included national, provincial, and local forestry officials, zoo representatives, and biologists. He warned forestry officials and local residents not to be tempted into exchanging pandas for cars, money, and radios with zoo officials.[60]

Han Zhengfu's pronouncements were not the first of their kind. A few months earlier, the national Ministry of Agriculture and Forestry tried to

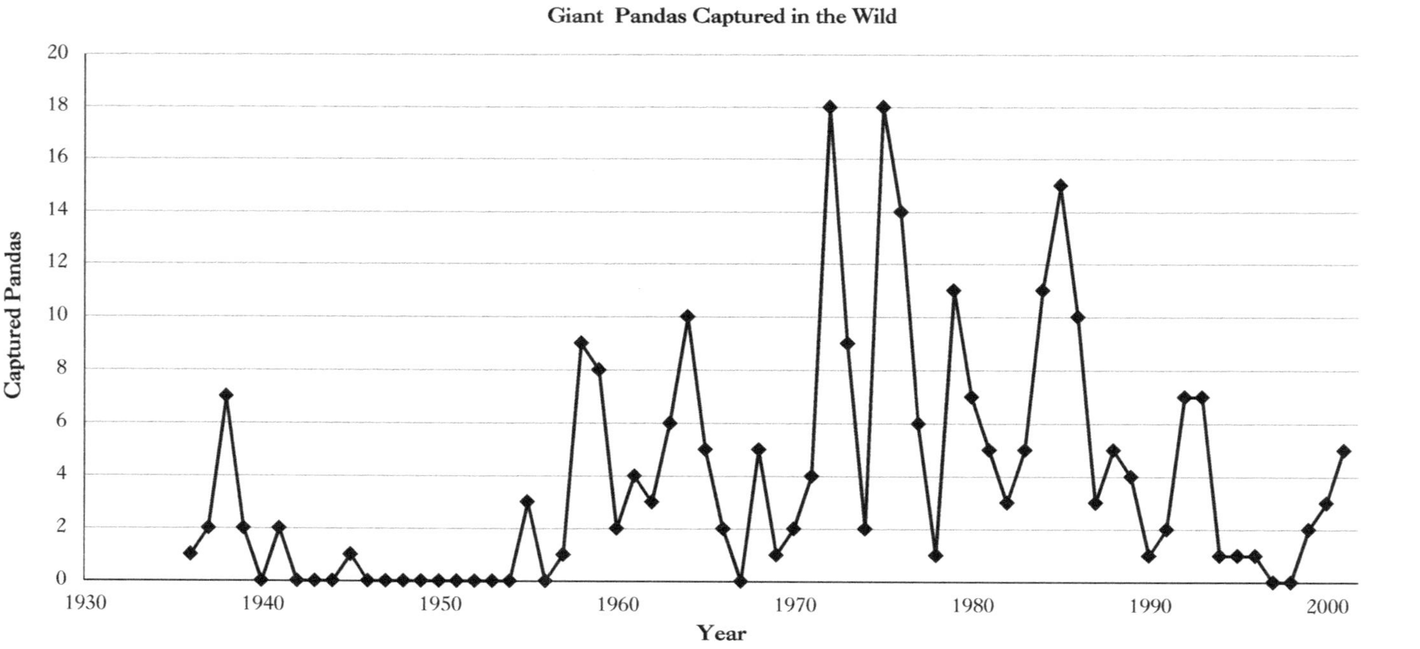

FIGURE 5.3 This graph illustrates the dramatic shifts that occurred in panda extraction from the wild. The highest rate of wild capture occurred during the 1970s to stock domestic zoos and supply foreign gifts during the peak of panda diplomacy. Other high-capture rates also occurred during panda-starvation scares, when many giant pandas were perceived as being rescued into captivity.

establish regulations on the proper uses of protected species and to limit this recent trend of excess.

> In this year alone [1973], zoos and research institutes in every district have requested our ministry to trap nearly two hundred fifty precious species of all kinds, but our nation's precious species are few in number, indeed treasured. Moreover, they have been subject to serious depletion and are not abundant. In order to strengthen protection, replenish these resources, and conduct research, we need to survey precious animal resources in all relevant areas. Please inform each zoo and research institute that in order to solve the present situation we must temporarily cease trapping.[61]

The Ministry of Agriculture and Forestry's effort to fight wild panda extraction was a difficult battle. All parties proved to benefit from panda trappings, apart from the pandas themselves.

These new policies and efforts to curb the trapping of wild pandas and other precious species achieved some limited success, at least initially. The number of wild pandas captured dropped dramatically for a couple of years. The high rate in 1976 can be attributed in part to a starvation scare from a widespread bamboo die-off, during which many pandas were "rescued" into captivity. The extraction of giant pandas from the wild grew again during the 1980s, but never again achieved the peak experienced during 1972.[62]

Concern over the damage that the panda's growing popularity was likely inflicting on the wild population also inspired further scientific study and protection of wild animals. In Sichuan, a province-wide meeting on surveying precious species produced several reports and suggestions for expanding animal protection.[63] The People's Government of Sichuan Province responded by initiating new panda and other animal surveys in 1974 that covered a much broader base than the 1967–1969 survey in the Wanglang reserve.[64] The content of the vast majority of nature protection initiatives that began appearing in 1973 was very similar to that of policies of the early 1960s. The interest in reviving these principles, including the concept of "protect, rear, hunt," indicated policymakers' desire to revive old projects and move forward with nature protection efforts.[65]

In spite of their intensified concerns about the recent escalation of precious animal trappings for zoo exhibits, forestry officials did not protest the practice of offering state-gift pandas. To the contrary, policies and speeches alike made explicit distinctions between typical, problematic zoo trapping and "appropriately" making pandas available for state gifts. Minutes from a national meeting on precious species protection praised Mao's use of the giant panda for international exchange, stating, "Precious animal resources are the treasured legacy of our natural history. The giant panda in particular, which has been called our 'national treasure,' has enabled our nation to make friends with the people of the world."[66] In his aforementioned speech to the Sichuan Province meeting on wild-animal protection, Han Zhengfu noted, "Chairman Mao's revolutionary foreign relations line serves a definite purpose, building friendships with the peoples of the world." [67] Forestry officials wanted to limit panda extraction to state gifts only and believed they needed to actively curb the heightened domestic demand that these high-profile gifts had inspired.

The transformation of the giant panda into a state gift added to its importance. Previous nature protection initiatives often emphasized that the giant panda was "precious and rare," "existed only in China," and was "important for scientific study." After the 1972 offerings of pandas to the United States and Japan, it was called "a tool of national and international importance" a means of "befriending the people of other nations" and "stimulating international exchanges and trade." This shift in rhetoric illustrated that protecting the species was not just important for science and the wilderness, but for the political wellbeing of China too.

Allowances for state-gift pandas became official policy. Item fourteen in the "1973 National Wild Animal Resource Protection Regulation" states:

> National first-, second-, and third-level protected animals without exception are not to be exported for trade. Only under necessary circumstances can a first-level, nationally-protected animal be given as a national gift, and only under the central government's control. Second- and third-level protected animals can, with the approval of managing authorized central-governmental ministries, be engaged in international cultural exchange.[68]

This provision, while allowing the central government to continue in its savvy international offerings, enabled the Ministry of Agriculture and Forestry to limit the export of all nineteen first-level protected precious species. In this way, wildlife protection advocates managed to seize upon the central government's state-gift policy and use it to create strict wildlife protection policies for China's precious species. Policymakers were also depending on the central government to be politically savvy in their gift offerings because bestowing giant pandas on too many countries would detract from their significance. Forestry officials thus succeeded in limiting the export of all precious species by emphasizing the political importance associated with one.

Limiting the authority to approve the export of the giant panda and other first-level protected species to the top officials of the Chinese government thwarted the ambitions of lower-level government and zoo officials to conduct panda trades. Attempts to circumvent this law by arranging other types of cultural exchanges for pandas took place in subsequent years. For example, in 1979 and 1980, two professors at the University of California, San Diego (UCSD), Paul G. Pickowicz and Wu Jiaowei, unaware of this 1973 policy, tried to arrange a giant panda exchange between the San Diego Zoo and the Chongqing Zoo. At that time, UCSD and the Chongqing Institute of Science and Technology had a sister-school relationship that entailed other academic exchanges. The San Diego Zoo, after being outbid by the National Zoo in Washington to host the state-gift pandas that resulted from Richard Nixon's visit to China, was anxious to obtain and exhibit giant pandas. Because there were several pandas at the Chongqing Zoo, the local officials in Chongqing were very enthusiastic about the idea of such an exchange. The proposal was rejected as soon as it reached provincial-level authorities, however, because of the regulations passed in 1973.[69] During the early 1980s, however, China ended its state-gift panda program, on the grounds that the animals were too precious to give away, and decided instead to institutionalize short-term loans of pandas.

State-gift pandas were among the most successful efforts by China show a new face as it strove for greater international recognition and integration during the early 1970s. Charismatic and lovable, the pandas were dramatically successful crowd pleasers in each country they entered. As such, these fuzzy creatures thawed Cold War tensions and promoted the idea that warmer relations with the inscrutable Communist power could be possible. China potentially had a great deal to offer in traditional state-to-state exchanges of strategic and economic friendship, but its most sought after offering quickly became its endearing black and white representatives.

6

Rescuing the Panda from a Reforming China

PANDAS BEGAN DYING, one by one, in the mountains of Pingwu County, Sichuan in the winter of 1975.[1] Lumber workers, local villagers, and forestry officials encountered their dead bodies during their course of work and errands that took them into the mountain forests.[2] Typically, dead pandas were spotted lying on their side in the grass or slouched in the snow, immobile and emaciated, with fur that was unusually coarse and lackluster.[3] The number of dead pandas reported was unusually high—too many to be the result of old age or an aggressive parasite. Other pandas seemed to be ailing—walking unsteadily, appearing out in the open, sometimes approaching villages. A Baima man brought a weak, disoriented panda to the county forestry bureau, which in turn delivered it to a zoo for care.[4] Once county officials realized that something unusual was happening in their forests, they began to send out investigation teams. Other counties were also affected, in Gansu Province as well as northern Sichuan, inspiring a survey team led by a local resident in southern Gansu to search the nearby forests for weak and dying pandas.[5] They encountered an adult panda that could still move but did not flee when they approached. The local guide pinched its ear, but the giant panda had neither the strength nor the will to resist.[6] Reports differ on exactly how many dead and ailing giant pandas were found in each area, but the authors of all reports and official correspondence on the matter agreed that they were amidst a serious crisis, an as yet unexplained mass death of the "national treasure." Such an event had not occurred in the history of the People's Republic, nor according to the scant records of pandas in historic Chinese texts any time in the past.

In 1983, the threat of another giant panda mass death emerged in the southern Sichuan range of the Qionglai Mountains. The threat was first noticed in the Wolong panda reserve, a larger and slightly newer reserve than the Wanglang reserve. At that time, George B. Schaller was directing an international panda behavioral study in the Wolong reserve organized by the World Wildlife Fund (WWF). Officials feared this second event was going to be a repeat of the first. As a result, the situation that created a crisis for the Ministry of Forestry and related departments in 1976 was turned into a national cause when it seemingly recurred in 1983. The juxtaposition of the ways that China dealt with the fear of mass panda death during these two eras illustrates the dramatic transformation that China, its society, its politics, and its economy underwent in just seven years.

Fatal Flowers, Episode I: 1976

The 1976 panda crisis occurred in a historic year of several monumental events. On January 8, Premier Zhou Enlai passed away and was subsequently mourned at an incredibly emotional state funeral. As early as February 10, 1976, the county officials in Pingwu were aware that something was amiss in the mountains that led to the panda deaths and filed an "urgent" report. The Mianyang prefecture and provincial-level offices replied within a month with further reports indicating that this calamity extended beyond Pingwu County. People were finding panda corpses throughout the Min mountain range in northern Sichuan and southern Gansu; Nanping, Beichuan, Qingchuan, and Wen counties all suffered losses of this prized animal in surprising numbers.[7] By March, urgent notices regarding the unusually high numbers of panda deaths flew between the county and provincial offices in Sichuan and Gansu and the national office of the Ministry of Forestry. The nation, however, was still in mourning for the deceased premier. In April, more than 100,000 people gathered at Tiananmen Square in Zhou Enlai's memory, an homage that turned into a protest. This expression of devotion to the deceased premier perturbed Cultural Revolution leaders still in power; they finally dispersed the crowds with warnings and force. During these early months of 1976, Chairman Mao became reclusive, at least in part because of ill health. Potential heirs to power began to jockey for position. By early July another leader, Zhu De, who had constructed and led the Red Army, died.

By this time the Ministry of Forestry had organized and sent the United Survey Team, a team of biology and forestry experts from multiple institutions, to investigate the widespread panda deaths in Sichuan and Gansu. On July 28, eight days after this survey team submitted its report and recommendations, the world's most devastating earthquake hit Tangshan, in northeast China, killing 242,000 and injuring 162,000 people.[8] In this broader context, the panda deaths were not China's biggest problem.[9]

About a month later, a series of three earthquakes hit the Min Mountains, the same place where the pandas were dying. Some speculated the earthquakes in the Min Mountains were related to the Tangshan earthquake, and many more viewed these various natural calamities as classical signs of the end of an era. During the dynastic periods, natural disasters such as earthquakes and floods, especially when several occurred consecutively, were seen as an indication from heaven that the emperor or the dynasty had lost the Mandate of Heaven to rule over China and a new era was about to begin. On September 9, 1976, Mao Zedong died. Within a month, the central government brought closure to the Cultural Revolution by arresting the Gang of Four, Mao's wife, Jiang Qing, Wang Hongwen, Yao Wenyuan, and Zhang Chunqiao. All the while survey teams collected dead and ailing pandas. The persistent attention to the pandas throughout this year—though it was filled with the loss of China's central leaders, political power shifts, and tragic natural disasters—was striking.

The many levels of government were first informed of the panda deaths by a series of rapid-fire reports: "Urgent notice on strengthening giant panda protection work"; "Urgent notice of the request for the collection of pandas who have died naturally"; "Urgent notice concerning the protection of giant pandas"; "Urgent notice concerning the survey of the giant panda situation"; and "Urgent notice on conducting giant panda protection work well."[10] If one thing is clear, it is that officials at various levels considered the sudden upsurge in giant panda deaths to be an urgent crisis.

Forestry experts did not take long to come up with an explanation for the mass deaths. The culprit was flowering bamboo. This is a natural part of the bamboo lifecycle, but because bamboo is the primary food for pandas, flowering can be problematic for them because it is followed by mass death of the stalks. The frequency of flowering varies by species and environmental factors, but flowering cycles tend to range from forty to eighty years. Because an entire hillside of bamboo may technically be a single plant, and even where there are multiple stands, a single species follows the same flowering cycle, large areas of bamboo died at the same time and turned expanses of a mountainside from green to brown. The giant pandas in the Min Mountains faced an additional challenge on top of starvation. Officials noted the unusually cold temperatures in the winter of 1975–1976. Some speculated that there might be a connection between temperature change and the bamboo flowering.[11] See Figure 6.1 for a map of bamboo flowering and die-off episodes. Officials expressed concern that the decreased temperatures would exacerbate the pandas' struggle to consume and maintain calories.[12]

The process of determining that bamboo flowering was the actual cause of panda death required further investigation however. Already in February of 1976, the Pingwu County Revolutionary Committee (the county government) encouraged local people to survey the areas and hand over panda corpses for autopsies.[13] The autopsies revealed that they died from a lack of nutrition.[14] A later study in Gansu Province offered more detail about how starvation transformed the body of the giant panda: "Their stomachs were empty or without food, there was no fat, the livers had assumed a grassy quality, not even a centimeter thick, and the abdomen was full of water."[15] Such images of the giant panda contrast with descriptions of the healthy young pandas at the Beijing Zoo in 1955 as fat with milk-white fur.[16] The national treasure was now ailing and emaciated. Something had

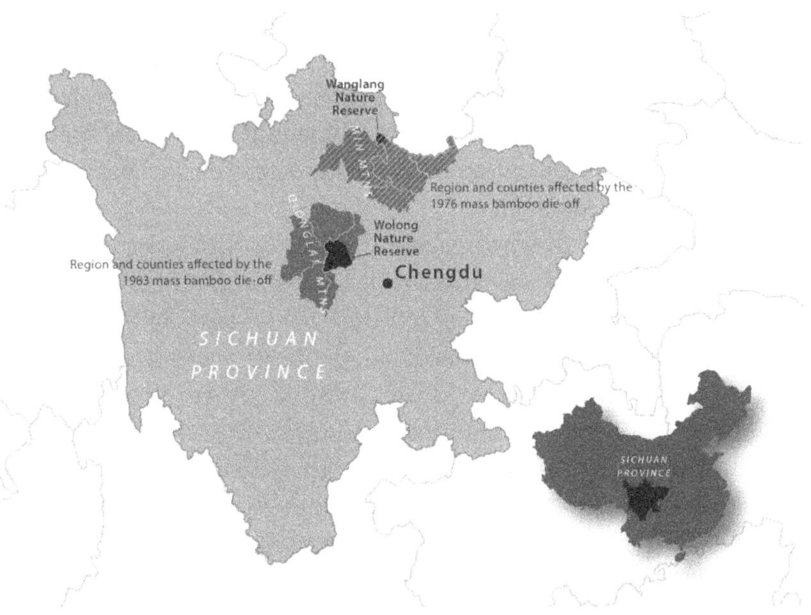

FIGURE 6.1 Both the 1976 and 1983 bamboo die-off episodes are displayed on this map illustrating the relationship between the major reserves and each of the affected areas. Map created by David Medieros, 2015.

to be done. The Ministry of Forestry's investment of money and extensive effort to try to ensure the continued existence of the giant panda, especially at the same time the PRC lost its top leaders and suffered one of history's most devastating earthquakes, demonstrated what a high priority the giant panda was for China's government.

Gathering the Sick and the Dead

Once the panda-starvation crisis and its cause were established, ameliorating it became a county-wide project everywhere the calamity hit. Each work unit was called upon to protect and rescue pandas. New regulations were also implemented to protect areas where bamboo grew and had not yet flowered.[17] Each level of government issued directives ordering the organization of teams to gather sick and dead pandas. Some of these directives were independent and redundant, while others indicated coordination and communication among various levels of government. The national forestry office responded with speed and force. After reiterating a call for the collection of the dead, the national office of forestry also commanded an investigation of bamboo flowering patterns and the creation of a team of experts. Although this new national survey team was clearly going to have to rely, at least to some degree, on the data collected from the county governments, there was no comprehensive plan for integrating and systematizing the various survey efforts. As depicted in Figure 6.2, rural peasants and Beijing scientific researchers united in rescuing the suffering giant pandas from the increasingly inhospitable environs in the Min Mountains.

By March 1976, with eleven pandas found dead, the Sichuan provincial government was highly concerned with getting to the root of the crisis. The Sichuan Department of Forestry alerted all relevant counties about the crisis and rescue measures. It called on provincial teams to survey other panda-inhabited counties and research ways to prevent the same crisis from occurring there.[18] In a subsequent urgent directive, the provincial forestry office asked county-level nature reserves and forestry bureaus to participate in the detailed survey with a team to be organized by the national government.[19] Although the giant panda does not exclusively live in Sichuan, it is home to the vast majority, so the provincial department of forestry considered itself the animal's main caretaker.

That month the Ministry of Agriculture and Forestry in Beijing responded to the panda-starvation crisis with resources. In Beijing, the

FIGURE 6.2 Local people rescuing a giant panda during the 1976 Min Mountain bamboo die-off. Photograph courtesy of Wanglang National Nature Reserve, Pingwu County, PRC.

central office of the Ministry of Agriculture and Forestry called on the major governing departments of the Sichuan and Gansu provincial offices, the Chinese Academy of Sciences, the research institute of the Beijing Zoo, the Botany Research Institute, the Beijing Natural History Museum, the Shaanxi Biological Resources Research Team, the Zhejiang Subtropical Forest Research Station, and others to select representatives to participate in a United Survey Team. Integrating talents from the different institutions and regions, the team set out for Gansu and Sichuan to examine the problem on the ground. The special team would make a more formal investigation into the causes and possible solutions to the problem, in addition to all of the local efforts already under way.[20] Saving the pandas warranted the rapid recruitment of representatives from several institutes for a prompt and thorough investigation. The giant panda was not merely a rare and exclusively native species; it had become a symbol of the People's Republic of China.

County forestry bureaus were the most active organizers of panda and bamboo rehabilitation and survey efforts. If dead pandas were found, their bodies were handed over to the county forestry bureau. Each body was registered with a number; the commune area in which it was found; a detailed description of its immediate location; the time; the general environment; and the panda's sex, size, and apparent reason for death.[21] If ailing pandas were found, experts and experienced hunters were sent to their location to feed or rescue them. Moreover, any unauthorized skinning or other use of panda corpses was forbidden.[22] Authorities wanted to ensure as safe and stress-free an environment as possible for the giant pandas, so the Pingwu County government temporarily banned hunting of any kind in any panda habitat, outside as well as inside the reserves. Hunting dogs were also forbidden in forest areas, because they were known to disturb, circle, and tree pandas. The county forestry bureau educated local people about all of the plans. It also offered compensation to people for expenses they incurred in their efforts to feed and care for pandas. Those who made notable achievements in the panda rescue effort were to be recognized and rewarded, while those who impeded these efforts could face withheld wages, forfeiting of holidays, and payment of fines.[23]

All levels of government investigating the panda-starvation crisis remained optimistic about preventing a future recurrence of this calamity. To this end, a two-tiered research agenda developed. Pingwu County began to survey its once abundant arrow bamboo and look for patterns. Bamboo surveyors examined bamboo and recorded the commune area, including a detailed description of its location, the mountain gorge name, and the type of bamboo. For bamboo that had already flowered, the time of flowering, the time of bamboo death, and the area it covered were all noted. For bamboo that had not yet flowered, the area, the maturity level of the bamboo and its present condition were considered relevant. The surveyors, shown in Figure 6.3, also investigated how long it had been since the bamboo had sent up shoots.[24] A lack of shoots was considered to be the first sign of flowering in some species.[25] They not only surveyed the hillsides and mountains but also perused volumes of traditional Chinese local histories called gazetteers for historic patterns.[26]

The Mianyang Prefecture level of the forestry department specifically called upon its lumber company workers to organize watches in areas where the bamboo was thin. The prefectural office overseeing the lumber company also called for research on bamboo and its recovery rate after the die-off, explicitly asserting that such knowledge could aid in future giant panda protection policy.[27] The active integration of hunters and lumber workers in the panda rescue efforts indicates an important shift in the perspectives of local and provincial governments from blame to dependence. Previously these two groups had been seen as engaged in activities that were incompatible with panda protection. Even if hunters did not illicitly hunt pandas, their activities often endangered pandas by accidentally trapping them in snares designed for deer or allowing excited hunting

FIGURE 6.3 Surveyors examining flowering bamboo in 1976. Photograph courtesy of Wanglang National Nature Reserve, Pingwu County, PRC.

dogs to harass them. Loggers destroyed panda habitat. Yet because both of these groups had intimate contact with the panda's environment and sometimes with pandas directly, they commanded specialized knowledge that could now help the panda. Biology experts and government officials alike recognized the importance of soliciting a wide spectrum of information and local knowledge.

The central government reiterated and encouraged further implementation of many policies already in place in the counties experiencing the flowering crisis, but it placed special emphasis on the collection of panda corpses for study and analysis. The obsession with collecting and cataloging expanded even to the point of maintaining a collection of panda skulls. The Ministry supported further hunting prohibitions and the punishment of those causing direct or indirect damage through reckless hunting or habitat destruction. Officials also promoted the elimination of wolves, leopards, or any other animals that might prey upon a panda weakened by hunger. While full-grown giant pandas are not normally at much risk from predators, anti-predator regulations had long been integrated in PRC hunting policies, and this measure reinforced a time-honored practice.[28] The central Ministry of Agriculture and Forestry used its authority to integrate a broad spectrum of actors throughout the crisis area in implementing this wide array of policies for nurturing and managing the recovery of the dead bamboo.[29] In addition to investigating the bamboo life cycle, the United Survey Team probed other possible factors that might have contributed to the timing of the flowering.[30] The Ministry of Agriculture and Forestry directed this effort toward future policymaking, as well as determining the potential length of the crisis at hand.

Another example of the government's investment in preserving the giant panda was the financial loss it was willing to endure. The fear of losing a significant portion of the giant panda population inspired the Ministry of Agriculture and Forestry to halt one of its main operations, logging, in a large area of important timberland. In 1976, Pingwu County alone produced 18,271,087 cubic meters of timber.[31] As pandas tend to live in forested areas, a significant percentage of this amount fell within the boundaries of giant panda habitat. Three of Pingwu County's main lumber companies were situated in the Baima River valley, the heart of panda habitat.[32] In 1976, Pingwu's Mianyang lumber company alone produced 2,567,600 *yuan* (approximately $320,950) worth of timber. The willingness of the central office of the ministry that directly oversees the timber industry to halt this valuable production demonstrated a strong commitment to the wellbeing of the giant panda.

Staff at the Wanglang Nature Reserve felt a particularly acute concern over the crisis, given its mission to protect giant pandas. As reported, by June of 1976 thirty-seven pandas had either been found dead or captured for rehabilitation within the boundary of the reserve. Reserve officials felt that the crisis demanded a detailed revamping of reserve management. Through such restructuring, they could "advance the establishment of National Socialism and implement Chairman Mao's foreign relations line, both causes being of high political and economic significance."[33] The claim of a connection to "Mao's foreign relations line" was more than casual rhetoric for the Wanglang reserve directors, since they participated directly in finding and trapping pandas that became state gifts.[34]

The reserve began its restructuring by building enclosures inside the reserve where they could nurse ailing pandas back to health. These were zoo like enclosures with indoor and outdoor areas for the animals. Figure 6.4 shows these enclosures still standing about 25 years later. This project, estimated to cost five thousand *yuan*, constituted one third of the reserve's construction budget proposal.[35] Wanglang also solicited the provincial food administration to assist with their panda rescue project since it would require more resources. At the time of this particular document, the staff at Wanglang was nurturing one giant panda for which they were requesting 55–60 *jin*[36] each of corn, rice, soybeans, and green beans per

FIGURE 6.4 This is a photograph taken in 2002 of enclosures built to tend to sick and starving pandas in the Wanglang Nature Reserve during the 1976 bamboo die-off episode. Photograph by E. Elena Songster, Wanglang Nature Reserve, Pingwu County, Sichuan, 2002.

month.[37] Even with provincial funds for the project, providing pandas with enough food to sustain them was not an easy task.

With many separate groups counting dead pandas, there were understandably conflicting estimates about how many deaths there actually were. In reports and publications on data collected through December of 1976, figures range from thirteen to twenty-three dead in Wanglang specifically. By some accounts, Pingwu County lost thirty pandas; others estimate double that. Estimates for the total number of pandas lost to starvation throughout the Min Mountains vary from 92 to 138.[38] A number of giant pandas ended up in zoos as a result of rescue efforts. Two separate county accounts claim that ten and twelve pandas were captured and sent to enclosures. A third analysis asserts that twenty-eight pandas were introduced to captivity from the entire region of the Min Mountains.[39] The 2001 captive panda studbook indicates that approximately fourteen pandas in various zoos were captured in the Min Mountains during the period of crisis.[40] The high number of twenty-eight may include pandas that were sent to Wanglang reserve and Pingwu County enclosures and were never placed in China's zoos.[41] If this was the case, they may have actually been returned to the wild or died in captivity. Another possibility is that they did not live long enough in captivity to be counted.

During the 1976 bamboo die-off crisis, many Chinese officials viewed captivity as preferable to death. Captivity, however, was not the only technique used to save giant pandas. Some were relocated to areas where the bamboo was not dying, although tracking or other means of determining whether or not this was a successful approach were not implemented. Statistics on released pandas are unfortunately not available. Some scientists feared that "rescue" was being used as a euphemism for "capture" well into the 1990s.[42] Few pandas were truly rescued and returned for a continued healthy existence in the wild.

Broad surveys of both pandas and bamboo dramatically increased the amount of data and knowledge about these animals, their eating behavior, and their position in this specific ecosystem. Fear of losing this special animal, or a large percentage of the likely limited population, motivated China's elaborate bureaucracy to react quickly and seriously. New policies and the recruitment of local farmers, hunters, and lumber workers heightened local awareness of the value of these pandas to the government. The government's response and the specific measures that were put in place set an important precedent for the next episode of blooming bamboo that threatened the food source of the giant panda. Precisely because of the significant loss of giant pandas in 1976, subsequently settled at 138, the government was on alert for future flowering and ready to act.

Fatal Flowers, Episode II: 1983

Less than a decade later, on another mountain range in Sichuan, a different species of bamboo began to flower. This concerned researchers who were aware of the crisis in northern Sichuan and southern Gansu in 1976. By 1983, however, China was a completely different country. In 1978, Deng Xiaoping had risen to power and began directing China through the "Four Modernizations," in agriculture, technological advancement, industry, and defense. This campaign incorporated increased foreign relations that included contracts with large foreign companies, the normalization of relations with the United States, and large-scale international research projects with such organizations as the World Wildlife Fund (WWF). A new sense of stability began to settle over China after decades of upheaval. This stability, combined with the growing economy, inspired a new strategy for addressing the growing concern over China's pandas.

The 1983 episode of bamboo flowering happened to occur within the boundaries of the Wolong Nature Reserve (see Figure 6.1) where George B. Schaller, then director of the wildlife conservation division of the New York Zoological Society, was directing an international cooperative study of wild panda behavior. He and a team of both Chinese and western researchers had been working in the area since 1980. Although Schaller was confident that the pandas in his specific research area would have ample food for the next year or two, depending on the flowering patterns of the bamboo, he was preparing himself for the worst. Schaller was aware of the behavior patterns of the local pandas and their preference for the specific type of bamboo that had begun to flower and die and expressed his own misgivings. "We were seriously concerned about the pandas elsewhere in these mountains. Little was known about the extent of flowering in other parts of the panda's range, though we assumed that arrow bamboo throughout the Qionglai Mountains had been affected. We had to prepare for a major effort to rescue starving pandas."[43] The Wolong Nature Reserve had, by some estimates, approximately two hundred pandas and, by others, closer to one hundred pandas.[44] The panda population in the Wolong reserve certainly could not withstand a devastating famine.

When the news of the 1983 mass flowering arrived in the offices of the central government, officials again responded promptly and assertively. In 1976, no one had been paying close attention to the flowering until several giant pandas were found dead, whereas in 1983 the flowering raised alarm before a single death.

The bamboo composition in the Qionglai Mountains is somewhat different than that of the Min Mountains. Several interviewees asserted that the mass flowering in 1976 was believed to be particularly devastating for the giant pandas because that area of the Min Mountains contained only one type of bamboo, arrow bamboo, so pandas did not have another species of bamboo to turn to when their primary food source dried up. Investigations by Schaller and his team revealed that the giant pandas were indeed out of luck during the 1976 bamboo die-off, not because the Min Mountains only produced one species of bamboo, but because multiple species of bamboo surprisingly flowered simultaneously.[45] The end result remained the same—no food. Two main types of bamboo grew in the Wolong reserve area. After conducting an investigation Schaller believed: "Since two or more bamboo species grow in most Wolong valleys, it is unlikely that the mass flowering of *Sinarundinaria* (a bamboo) there in 1983 will cause widespread starvation among pandas."[46]

Regarding the potential threat to the giant panda as very serious, however, the Ministry of Forestry did not want to take any chances. Instead, the central government took action. The government mobilized old investigation and rescue plans and organized new groups for the job. This time, instead of organizing groups to gather the bodies of dead pandas and rescue ailing starving pandas, groups were organized to look for any indication or evidence that the pandas might be in danger of starving.

Seeking the Sick and the Dead

The Ministry of Forestry reinstated many policies that were part of the response to the earlier flowering incident, including the construction of panda holding stations, rescue committees, surveying teams, hunting restrictions, and rescue rewards.[47] Unlike the earlier bamboo die-off, the threat to the giant pandas became a nationwide public cause. While the 1976 flowering did not appear once in the *Renmin ribao* (People's Daily), more than thirty articles on the flowering and rescue efforts appeared in this same paper between 1983 and 1984.[48] The 1983 bamboo flowering permeated every form of media, and the public rallied with sympathy, nationalist sentiment, and monetary contributions before a single panda had even died.

People from all levels of government and society mobilized to address this potential problem. High-, middle-, and low-level officials held planning and strategy meetings. In August of 1983, the provincial government of Sichuan gathered in Chengdu, the provincial capital, to discuss the issue of starving pandas, organize small planning groups, and establish special measures for panda rescue.[49] The national State Council also sent a directive to the Ministry of Forestry urging aggressive and diligent action, particularly by relevant departments in affected areas.[50] In September, the Ministry of Forestry met to decide on what course of action to take and what new special measures to implement.[51] The provincial governments of Sichuan and Shaanxi Provinces held an urgent meeting in October for county and other local leaders on saving the giant panda. The meeting was also designed to guide these leaders in carrying out the spirit of the State Council's earlier initiative on the protection of wild animal resources.[52] These meetings resulted in multi-layered efforts to publicize their concern about the pending crisis, mobilize people into action, and set up coordinated efforts to rescue pandas.

Panda rescue measures were both recycled and new. For one, there was an increased effort to manage hunting and fight forest fires. Again, the State Council wanted to find out exactly which species of bamboo were flowering and the degree to which the pandas were threatened in each area.[53] Increasing scientific knowledge related to the wild panda's circumstances was a primary means of responding to this issue. In the provinces of Sichuan and Shaanxi, panda rescue workers participated in a short-term training course to improve their technical skills and to better understand the habits and characteristics of pandas and the physiological characteristics of bamboo.[54] Money later raised by American school children to save the panda was dedicated to "study alternate feeding programs for the pandas."[55] Other response plans included "placing the starving pandas in specially constructed recovery stations and later transferring the pandas, one at a time, to [a] new habitat."[56] These rescue techniques were front-page news in the *Renmin ribao*, even without a specific rescue to discuss.[57]

Sichuan Province also instigated the "Three Prohibitions" (三不准) and "Preventing Four Means of Death" (四不死) campaigns. The Three Prohibitions included bans on cutting bamboo or harvesting bamboo shoots, trapping or firearm hunting, or releasing hunting dogs into panda habitat.[58] The four causes of panda death that the rescue workers endeavored to prevent were forest fire, hunger, freezing, and hunting. Similar restrictions had been enacted during the 1975–1976 bamboo die-off, but without the media-enhanced catchy campaign slogans, such new policies had not attracted widespread attention. Increasing specific restrictions did little, however, if they were not enforced. Hu Jinchu, who had conducted research on giant pandas in the wild for several years and had become a recognized expert worldwide, favored greater enforcement over more restrictions. While working with George Schaller in the Wolong reserve, Hu Jinchu had witnessed pervasive problems with trapping within the reserve and therefore advocated making punishments stricter so that people would be dissuaded from hurting or killing pandas.[59] Hu asserted that "Even though panda territory is difficult to traverse, patrollers must penetrate deeply into the adverse environment if they are going to have any hope of finding any sick and hungry pandas."[60] Regardless of how many surveyors and panda scouts were hired, their efforts would be futile if the workers were not ready to take on this arduous task. Mobilization of local people had been noteworthy during the first panda crisis in the 1970s, but earlier recruitment efforts did not compare to the extensive canvassing and assembly inspired by the 1980s campaigns. With greater coordination, the three panda-inhabited provinces organized three hundred people into fifty-six patrol teams. Another 1,150 people participated in other surveys.[61] The extent to which those mobilized in rescue efforts embraced Hu's standards is impossible to know, but local recruits certainly were familiar with the landscape.

Another popularly discussed and quickly implemented policy stemmed from the notion that pandas are vegetarians that actually like to eat meat. Because giant pandas indicated a taste for meat in captivity, the Schaller research team used meat and bones to lure them into their traps so that they could put radio collars on them. Rescue workers in the Wolong Nature Reserve, who were not associated with Schaller's research group, placed approximately 4,000 pounds of beef, lamb, and bones on the hillsides to draw the pandas down to lower altitudes where there was an abundance of another species of bamboo not in bloom.[62] Although meat had proven to be a successful bait, Schaller expressed skepticism about this particular lure and rescue technique, noting: "I felt that pandas would no doubt descend on their own when they were ready as they had done in the past, and that few would find such bones. But weasels, martens, and other small creatures would delight in the bonus of meat and scraps."[63]

Once the various rescue efforts were enacted, officials met to assess their efficacy. Representatives from all panda-inhabited counties, regardless of whether or not the bamboo in their particular county was blooming, gathered in Chengdu during early March of 1984 to participate in the "Saving the Giant Panda Work Report Meeting."[64] Independently zoologists also held a special session on the panda-starvation issue during their annual national zoological meeting in Nanjing. There experts from around the nation discussed the "present difficulties of the 'national treasure' and how to better protect the giant panda."[65]

The first dead panda that was found was actually in the Min Mountains of Pingwu, outside of the area where the bamboo was flowering.[66] This discovery fueled concerns, even though it was unrelated to the mass flowering. Dead pandas were evidence of crisis and need. While every dead panda was a tragedy, each death also served to justify the investment and concern that had been generated so far and the solicitation of more funds and support. Panda deaths were broadly publicized and, according to some, fictitiously multiplied.[67]

Newspapers and officials both saw the benefits of maintaining a concerned public. In October of 1983, the *Renmin ribao* noted that, while no pandas had yet died in the Wolong reserve, about eighty were threatened by the bamboo die-off.[68] As reported in the United States, four pandas had died of starvation by February 1984, eight months after the first flowers were detected.[69] In July, the *New York Times* reported that only one panda within the reserve had died of starvation. This figure was later corrected to zero in a journal article by western and Chinese researchers who were studying in the reserve at the time.[70] According to a March 1984 article in the *Renmin ribao*, fourteen pandas sick from hunger were reportedly rescued in the three panda provinces. Nine of these pandas survived and seven others were found dead in the wild, bringing the death count to twelve.[71] In April, the *Renmin ribao* conducted a special interview with the head of China's Wild Animal Protection Association in order to update the newspaper readership on the panda situation. The head of the association did not provide any death statistics, but rather noted that "Finding sick and hungry pandas is not easy. Rescue work is, therefore, a long-term task." Later in the interview, he asserted that "in this period of time rescue work has been very successful, we've rescued several ailing and hungry pandas."[72] By another account in September 1984, fourteen pandas had been rescued, five had died during rescue efforts, and four dead pandas had been found in the wild, bringing the death count to nine, three fewer than the March accounting.[73] Panda death reports were inconsistent, but all reports stressed the threat that China's giant pandas faced.

Once some sick and dead pandas were found, panda rescue stories became a favorite press item in China and internationally. Whether moved by personal concern for the giant panda or monetary awards, a number of people who lived in the vicinity of giant panda habitat gained fame through heroic rescue efforts. Local people were often glorified in newspaper reports for their remarkable efforts, including material sacrifices that the individuals, usually very rural peasants, were reported to have made on the panda's behalf. One person notified the county government of Tianquan County that a giant panda, presumably weak and disoriented from hunger, had fallen off a cliff. The veterinarians at the Chengdu Zoo worked to save the animal.[74] This story was retold in a *New York Times* article with the added detail that this peasant and his neighbors "carried the panda on a light tractor to the county seat 50 miles away in a 19-hour trek. The peasants received a $460 reward."[75] This amount was more likely 460 *yuan* (about $58), which at that time was approximately equal to the average yearly income for a person in the mountain counties of Sichuan.[76] Qin Yiling, a peasant of Pingwu County, brought a panda that had wandered into his house presumably looking for food to a government holding station. In Lushan County, a panda reportedly killed a farmer's goat, but the farmer not only refrained from shooting the panda, but also restrained his dogs inside the house so that they would not harass the panda.[77] The vice-chair of the Ministry of Forestry, Dong Zhiyong, recognized the great restraint with which this man acted in the face of his economic loss and commended him for it. In another area, a group of local people actually slaughtered one of their own goats to nourish a panda, which they subsequently guided to a healthy bamboo grove. The loss of a goat in the near subsistence economy of most mountain villages of Sichuan was a significant economic sacrifice, but if peasants could provide evidence or engage an official in the rescue effort, they could gain substantial monetary awards by aiding pandas. The financial rewards would certainly compensate them for gas, a goat, or other expenses they incurred in the process of their rescue efforts. These stories most often emphasized the financial sacrifices of the peasants over the rewards, which were depicted as a benevolent government's expression of appreciation.

"Love Your Nation, Love the National Treasure, Save the Giant Panda"

In addition to inspiring many strategic planning meetings and rescue missions, the looming panda-starvation crisis inspired many substantial donations, domestic and international. After the Chinese government publicized that it had established a special fund to save the giant pandas, money began to flow in from all directions.[78] In January 1984, the Wild Animal Protection Association—a subsidiary organization under the Ministry of Forestry established to address concerns related to precious and rare species and to promote scientific research and captive breeding—decided to devote that year's annual international fundraising efforts to saving the giant panda. The organization solicited donations from Chinese airports, hotels, and zoos. In addition, it organized fundraising activities through wild animal and nature reserve organizations in Japan, Australia, Canada, and the United States.[79] Foreigners living in China expressed their concern for the giant panda by pouring funds into the purse of the Wild Animal Protection Association. Additionally, each of the forty embassy workers at the West German embassy donated funds to the Ministry of Forestry for panda saving efforts, asserting that the giant panda is Germany's "favorite precious species."[80] A Chinese foreign student in

Sweden presented China with 3,000 Swedish crowns for the pandas.[81] Money also arrived from other countries, including Hong Kong, Japan, the United States, West Germany, and overseas Chinese communities from all over the globe.[82]

Concerned citizens inundated the *Renmin ribao* with letters inquiring about the status of the panda-starvation crisis, rescue efforts, and level of donations to date. In response to popular demand, China's main newspaper interviewed officials in the Chinese Wild Animal Protection Association. Fundraising was one of the primary points of interest for readers. Without accounting for all of the money that was channeled into the Ministry of Forestry or the money that went straight to local governments in panda territory, by April 1984, the Wild Animal Protection Agency had received more than 400,000 *yuan* from domestic donations (approximately US $50,000), about US $40,000 from embassies and tourist organizations, US $200,000 from a Japanese television station, and twenty-one trucks from the Japan office of the World Wildlife Fund.[83] These twenty-one trucks and US $290,000 represented just a fraction of the funds that the fear of starving pandas inspired. The World Wildlife Fund donated an additional US $200,000 toward panda rescue efforts to ensure that its international wild panda behavioral research project, already under way, would not be derailed by this potential crisis.[84] Even after US $490,000 were raised, contributions continued to grow.

One of the most famous international contributions came by way of Nancy Reagan. The United States World Wildlife Fund ran a campaign called "Pennies for Pandas," which encouraged American schoolchildren to donate money to save the pandas in China. After the schoolchildren raised US $13,000, they presented a check to First Lady Nancy Reagan to take with her when she accompanied the president to China.[85] Several hundred Chinese children, all waving panda flags, greeted the First Lady at the Beijing Zoo. Nancy Reagan presented a check to the head of the zoo and told the children present that the donation was from children in the United States to help in their efforts to save the giant panda.[86] This act solicited a hearty thank you from Deng Xiaoping. His first comment to Nancy Reagan when they met in Beijing was, "You have done a great deal for our giant pandas, thank you." Reagan accepted Deng's gesture of gratitude but admitted that she had received a great deal of help from American schoolchildren. The giant panda once again became an integral component of diplomatic relations between the United States and China.[87]

While affection and sympathy inspired foreign contributions, nationalist sentiment prompted extensive donations from Chinese citizens. Railroad-working teenagers were front-page news when they decided to organize and raise money for the pandas. Teenagers on the Datong railway line took the initiative and encouraged all of the young people associated with the national railroads, numbering more than 2 million, to participate in a fundraising campaign to help the giant panda "endure this famine." This particular fundraising campaign entreated railroad youth to "Love your nation, love the national treasure, save the giant panda" ("爱祖国，爱国宝，抢救大熊猫").[88] The term *guobao* (国宝), or "national treasure," came into heavy use during the 1980s, particularly in relation to the broad campaign to save the giant panda, and soon became synonymous with the giant panda.

Shanghai schoolchildren developed an elaborate donation system to protect the "national treasure" (国宝). The Number Ten middle school recruited 530,000 students to pledge one *fen* (a Chinese penny) for each day of their winter holiday to the cause of saving the "national treasure." The *Renmin ribao* (People's Daily) also featured images like Figure 6.5 showing children using stuffed animal pandas to help them raise money for the adored animals. Each student wrote her or his name in a pledge book to be sent with the funds collected for panda rescue efforts.[89]

The Chengdu army academy, inspired by the newspaper articles and radio and TV broadcasts, initiated a collection program under the slogan "Save the 'national treasure,' Love your country, Share for the sake of the nation" ("抢救'国宝'， 热爱祖国， 为国分忧"). In a matter of months, they had collected 2,541 *yuan* and 1,270 pounds worth of grain in food tickets.[90] A middle school in Chengdu devoted a schoolwide assembly to the plight of the giant panda. A biology researcher and activist lectured to the school about the giant panda's special biological traits and the importance of saving the giant panda. The school's students, teachers, and workers "transformed their nationalist spirit into action and enthusiastically threw their all into 'saving the national treasure (the giant panda)' activities. Some teachers contributed their [magazine, etc.] writing income; others contributed a full month's salary. Within a few days, they raised more than 500 *yuan*."[91] Even though Chengdu was just a few hours away from panda territory and the reported flowering, local residents expressed their concern for the giant panda in the spirit of nationalism rather than in terms of localism. Similarly, Chinese citizens from all over the nation contributed, even those who had little to scrape together. These efforts

FIGURE 6.5 People young and old organized fundraising events of all kinds to raise money to save the giant panda. This photograph from the *Renmin ribao* (People's Daily) shows the concerted efforts of children in China to raise money. The megaphone demonstrates an organized committed effort that was also endorsed by adults. The stuffed animals reflect a particular type of affection children harbored for these animals and an effort to elicit that affection among potential donors. Source: *Renmin ribao*, May 9, 1984.

expressed concretely a broad sense of ownership of, duty toward, and identity with this "national treasure."

Officials in charge of the panda rescue effort were impressed by the emotional and material outflow from the Chinese people. China's Wild Animal Protection Association was receiving a daily plethora of letters and donations from all over the country and the world. People from all walks of life—workers, peasant farmers, members of the military, and elementary, secondary, and university students—sent in monetary contributions. Dong Zhiyong, the vice-chair of the Ministry of Forestry, was particularly moved when all of the inmates at a labor camp in Jiangsu Province in Southeast China dipped into their meager stashes and collected 180 *yuan* (about US $22) to donate. The money was enclosed in a letter that read, "We offer this donation to help the giant panda and as an expression of our deep love for our nation. With this concrete action we are correcting our mistakes and walking in a new direction."[92]

Biologist Hu Jinchu saw the broader potential benefit to the nationalist sentiment that the ailing giant panda was inspiring. Beyond this particular panda-starvation fear, he sought to use this event to educate people about nature more generally and about the need for a long-range perspective on nature protection. He encouraged readers look at the larger problem of habitat destruction. Hu Jinchu advocated lectures and new curriculum at elementary schools to educate the future generations. "Teach them to love their home towns, to love the nation's mountains, waters, grass, and trees, to love nature, and to love science in order to make the nation thrive and prosper, and for these reasons to strive to contribute substantially to the nation."[93]

Perpetuating the Panda Plight

No one likely anticipated the public response would be so heartfelt and the material contributions so abundant. But, if pandas were not dying at a crisis rate and money was not all spent on the rescue project, then funds and attention from the press were benefiting more than just the pandas. As the reports on exuberant fundraising multiplied, people began asking more questions about how the funds were being used. Newspapers and magazines began turning to experts for information on the progress of the crisis. Panda fund assessments were highly inconsistent. There was some effort to centralize the donations through the central office of the Ministry of Forestry and redistribute them as needed. Inconsistencies in donation reports stem in part from the fact that some donations were directed to the Ministry of Forestry, while others went straight to the Wildlife Protection Association or provincial and local governments. Regardless of the details about how much money was raised or how many pandas died, there was a common narrative that everyone from the media and government to the general public found satisfying, namely that although the pandas had been suffering from starvation, the government was successfully fighting the crisis, thanks to the extraordinary funds raised by people from all walks of life.

Some see the 1983 bamboo flowering turned panda-starvation crisis as a story of much ill-gained and misspent money, but it had seemingly innocent origins. With a worrying number of pandas having perished in 1976, there was real reason for concern and a need for raising money and acting quickly to prevent a recurrence. The incredible influx of money was an unanticipated benefit. Those in the forestry departments and local governments who were supposed to be stewards for the intended beneficiaries found themselves with more money than they could invest in a seeming crisis that, with the exception of a few areas, was really more of a threat. As a result, corruption, or at the very least, irresponsibility, ensued. Instead of putting a stop to fundraising, officials, with the help of the press, perpetuated the impression that more money was needed.

One week before Nancy Reagan arrived for her visit in China, a front-page news article in the *Renmin ribao* explicitly stated that 80 percent of the giant pandas in Wolong Nature Reserve had moved to a lower elevation and had begun eating a different species of bamboo. Yet the move to more fruitful eating grounds, according to this article, did not indicate that the pandas were no longer in danger. One article, entitled "Because the arrow bamboo has mass-flowered and died, in Wolong giant pandas have no choice but to descend to eat the umbrella bamboo," gave no indication that this was actually good news. The article also quoted the scientific experts working in the Wolong Nature Reserve as saying, "In this area lies a latent threat of food shortage for giant pandas and within the next ten years rescue and protection efforts will remain a formidable task."[94] News about the giant pandas was consistently framed as point of grave concern.

Officials in the Ministry of Forestry and its subsidiary departments also consistently assessed the situation as a panda-starvation crisis. In July of 1984, according to a *New York Times* article, the head of the Sichuan provincial branch of the Wildlife Protection Association asserted that between 90 and 95 percent of the arrow bamboo in panda habitat was dying. The article did not indicate that this was only in some areas or that other bamboo species existed in many of the areas. After explaining the various measures that China had taken to protect the giant pandas from the bamboo shortage, Hu Tieqing was quoted as saying that if the Chinese government could completely finance the rescue efforts, their department would keep US $96,000 worth of donations in a trust fund "for the next generation." According to the article, he then solicited more contributions from overseas. Embracing a seemingly global spirit that in some ways countered the deeply-felt nationalist spirit inspiring so many Chinese donations, Hu Tieqing said, "the giant panda belongs not just to the Chinese, but to the whole world, so the whole world has an obligation to help the pandas when they are threatened."[95] In the context of a closely monitored press, it is notable that the author of this article raised doubt about the delegation of donations. Regardless of whether funds were irresponsibly delegated, unreported, or redirected, references to placing donations in bank accounts and trust funds,[96] even if fiscally responsible, belied the notion that funds were urgently needed for a crisis situation.

Finally, in December 1984, a centralized panda count estimated that thirty-three giant pandas were found dead in the wild, and thirty were rescued. Of those rescued, twenty-one survived, thirteen of which were returned to the wild.[97] According to this source, the final panda death toll was forty-two, including those found dead and those that died in the rescue process. These numbers also imply that eight rescued pandas were placed in captivity. The giant panda studbook lists fourteen pandas captured from the wild and placed in captivity during this time period.[98] Perhaps not all of these panda captures were considered to be rescue operations. Significantly higher than the September figures, the December statistics reflect an effort to synthesize data from several sources and are generally regarded as the most comprehensive. Since there was such a discrepancy between figure reports on dead and captured pandas, it is difficult to assess the accuracy of the final data. Yet even at this point the pandas were still being portrayed as in crisis. Vice-Chair Dong Zhiyong stated in this same December report that the panda crisis was still in full swing, nineteen months after researchers first detected flowering bamboo in Wolong. While noting that panda protection and rescue efforts had been very successful, his underlying message was: "The crisis has not yet subsided, in some areas the flowering bamboo continues to expand. Protecting and rescuing the giant panda is, therefore, a long-term task."[99]

Assessing the actual threat of the 1983 flowering also proved to be a long-term task. In 1988, Kenneth G. Johnson, George B. Schaller, and Hu Jinchu co-authored a paper arguing that no pandas in the Wolong Nature Reserve died from starvation directly.[100] The authors based this conclusion on data collected from 1981, two years before the bamboo die-off, through 1985, two years after the die-off. They measured bamboo consumption and grazing habits, range activity, and reproduction behavior on radio-collared giant pandas. They synthesized this data with analysis of feces and bamboo stands.[101] This was not to say that pandas did not die during the feared crisis or that none died of starvation outside of the Wolong Nature Reserve. In fact, the paper explicitly points out that the "die-off caused starvation locally in several areas of the Qionglai Mountains where alternative food sources were no longer available."[102] This second conclusion was not based on a separate study, but rather independent knowledge of the bamboo distribution and pandas in other sites. Rescue efforts did continue for a few years after the bamboo die-off, as did the continued collection of panda corpses. According to the statistics available to Johnson and others, sixty-two pandas died in the wild between 1983 and 1987. Twelve more died during the course of rescue efforts, bringing the total to seventy-four. An additional twenty-seven were captured because they appeared ill, twenty-five of which were re-released to the wild.[103]

Pan Wenshi, a wildlife biologist who worked with George Schaller in Wolong, uses both the data collected by Johnson and others and those cited by Dong Zhiyong in December 1984 to argue that the flowering of bamboo was unlikely to have threatened the lives of giant pandas. He does not rule out the possibility that the extreme case in the Min Mountains of many different species of bamboo simultaneously flowering could have resulted in a panda-starvation crisis. Even while refraining from a definitive judgment on the Min Mountains flowering, Pan remains skeptical that every panda that died during the 1970s flowering period died because of starvation, directly or indirectly.[104] He argues that in a wild population of approximately 1000 individual giant pandas, as estimated by the 1989 giant panda survey, an average natural mortality rate should be approximately fifty pandas per year, placing the death statistics from both the Ministry of Forestry and Johnson within the range of normal natural mortality.[105] Pan, therefore, does not consider the 1983 bamboo flowering and die-off a crisis. In his own parallel study of giant pandas in Shaanxi Province (1984–1998), he concluded that even while there was mass flowering of bamboo in his study area in 1983 and 1984, not a single panda died from starvation over these fourteen years.[106] In 1983, Pan Wenshi objected to the glossy press, widespread concern, and expansive fundraising that surrounded the bamboo flowering detected in Wolong.[107] He claimed that his objections fell on deaf ears.

Because the 1983 panda-starvation crisis grew out of genuine concern based on historic events, the government decided to learn from history and act preemptively in order to nip this second potential calamity in the bud. Starting out as some circumscribed misgivings that experts brought to the attention of the government, concern for the panda quickly expanded into a national and then international cause to save it. The young, the old, the crippled, and the convicted were each scrounging to offer even minuscule financial support for the cause. More substantial funds poured in from the government and concerned international citizens. Media groups donated significant amounts while seeking rights to cover the effort. Government officials gratefully solicited more contributions and promoted publicity that fed on sympathetic and nationalistic emotions and fear among the public. While some suspected this at the time, investigations conducted over subsequent years conclusively demonstrated that an emergency never really existed, at least not to the degree that it was portrayed. The bamboo situation was certainly worth monitoring, but ultimately did not truly endanger China's pandas.

Because a year of campaigning and rallying support could not be undone and lunch money could not be returned to elementary school students, the matter was quietly closed. A few more news articles on the great success of the panda-saving campaign calmed the concerned public and assured the donors.[108] Corrections popped up in foreign specialized journals and a few domestic books. The panda-starvation scare thus never became scandal. Instead it is remembered as a rallying point for the nation and a crisis resolved. The greatest benefit of this episode was that it inspired some talented youth to later pursue careers in conservation biology and join the ranks of China's top panda specialists.

The giant panda came to life most vividly when it was faced with death. Even when China was seemingly collapsing with the fall of its top leaders, Zhou Enlai, Zhu De, and Mao Zedong, and the tragedy of a demolished city, the ministry charged with protecting the giant panda responded with full force to the urgent call to save the "national treasure." Amidst privatization and rising standards of living, the Chinese people handed over their long-awaited, newly gained income for the panda. The fear of losing the panda to these flowering episodes also motivated people to resist the forces of their times, be they fatalism or self-aggrandizement, and focus on rescuing this unusual animal from flowers, death, and extinction. In so doing, China's citizens were saving themselves from a China without the panda, an unthinkable outcome when the animal had become so completely entwined with their national image.

7

Coloring the Panda with an Ethnic Touch

MONITORING PANDAS AND ECOTOURISM

This is a trip for nature enthusiasts, who revel in meadows of wildflowers, enjoy dining on fresh wild mushrooms, and like to experience and learn about nature on a holiday adventure. A visit to a Baima village promises much fun as colorfully dressed young women serenade guests with traditional song, and serve an excellent meal of barbequed goat, wild vegetables and homemade honey wine.[1]

THE ALLURING DESCRIPTION of an ecotourism excursion to the Wanglang Reserve reveals a dramatic transformation in the ways that people perceived the habitat of wild pandas just one decade after the panda starvation scare of the 1980s was put to rest. Beginning in the late 1990s, ecotourism became a promising panacea for the perennial struggles of panda reserves; it was seen as a new means of supporting the protection of pandas and other endangered species, especially for those species that live in beautiful landscapes. As whimsical and romantic as this ecotourism description appears, the concept of ecotourism was a calculated program proposal that resulted from long deliberations among scientists about ecological concerns. Such basic principles as the recognition that people who live in and around nature reserves also need a form of livelihood and that lumber trade devastates panda habitat formed the underpinnings for the introduction of ecotourism in panda country. The plethora of complicating variables that emerged as ecotourism evolved at Wanglang, and more broadly in China, revealed the true complexity of this new industry. Scientists, NGO representatives, reserve staff, and local residents contributed to redesigning Wanglang with an eye to transforming it into a model ecotourism site. Scientists and reserve staff originally envisioned ecotourism as a complement to the core scientific research and conservation work being done in the reserve, but it stole the limelight, at least for a while. Later, as ecotourism took on a life of its own, other forms of scientific research and conservation studies proved to be much more sustainable activities in the reserve. Ultimately, Wanglang's true pioneering legacy is found in the reserve's persistent commitment to scientific research and innovation in conservation.

The emergence of the Wanglang Reserve as a model research base and innovator in wildlife conservation programs for nature reserves in China is a concurrent narrative that repeatedly intersects with the story of converting Wanglang into a model ecotourism site. Both identities reflect Wanglang's efforts to adapt its role as a giant panda reserve to the rapidly transforming political, economic, and social landscapes of China at the turn of the twenty-first century. As Wanglang evolved into a scientific research base, two of the most long-sustaining conservation projects include an increasingly sophisticated wildlife monitoring program and coordinated staff training. These efforts proceeded steadily in the background while ecotourism to Wanglang experienced dramatic and sometimes unexpected shifts in different directions.

Ecotourism emerged during the late 1990s in part as a response to the impacts of the 1980s Reform-era policy changes. The nation's focus on privatization spread across China's mountains with the dramatic expansion of logging, which quickly exacerbated the degradation of panda habitat. In spite of the panda starvation scare of the early 1980s and the mass fundraising campaigns that accompanied it, Wanglang and many other remote panda reserves obtained only minimal funds with which to function. Pingwu presented itself as a lonely county in northern Sichuan province that seemed to attract more attention from pandas than people. This was reflected in subsequent panda population surveys that showed a relative abundance of pandas in this particular county.[2] Although the political discontent that erupted in Beijing at the end of the decade and culminated in the tragic Tiananmen massacre had little direct impact in China's remote mountain landscapes, the dramatic uptick in economic growth that Deng Xiaoping ushered in after of the Tiananmen debacle did reach China's high-elevation hinterlands. Accelerated economic expansion beginning during the early 1990s introduced intensified human activity in panda country.

When scientists and the WWF discovered that Pingwu County was home to more pandas than all of China's other panda counties, the area became a priority for panda protection efforts.

Trading Timber for Tourism

The Baima people who lived around the Wanglang Reserve were agriculturalists in a marginally productive environment who historically struggled for sustenance. Aside from their truncated dealings with opium cultivation prior to the founding of the PRC, they cultivated no crops that were particularly lucrative. Timber companies were previously state owned and operated, so the Baima were largely excluded from the fruits of early timber harvesting.[3] Furthermore, the trees were considered national property, so when logging activity expanded during the 1950s, the state logging company stripped timber from the hills surrounding the Baima villages.[4] During the 1980s, management of logging, like that of most other industries, was shifted substantially to the household level. Many Baima people pooled money and purchased trucks in order to haul lumber, and as state lumber companies extended harvesting rights to a broader group of participating households, Baima became increasingly involved in and dependent upon the lumber industry for income.[5] During the 1960s, the Pingwu government had berated the Baima for hunting because it threatened the giant panda and other recently designated precious species.[6] With the integration of the lumber industry into the Baima economy, Baima subsistence once again was on the wrong side of nature protection. This time, however, it was not the county government that expressed concern, but an international non-governmental organization (NGO), the World Wildlife Fund (WWF).

The WWF was working in China as a result of Deng Xiaoping's "open door" policy, which he launched in 1978.[7] One rationale behind this policy was that China could benefit from international expertise by inviting foreign experts to share their knowledge and technology with the people of China. Prior to this period, Mao Zedong had promoted extensive exchanges with other socialist and developing countries, but not with the industrialized West. This policy was yet another component of China's post-Cultural Revolutionary scramble to make up for the decade of stunted education and economic development. The WWF-sponsored cooperative wild panda behavioral study under George Schaller's leadership was one of the early material manifestations of this "open-door" invitation to foreign experts.

The WWF was invested in protecting endangered species broadly, but was particularly interested in the preservation of the panda because it was the featured animal of the WWF logo. With the guidance of Dr. John Mackinnon, the WWF gathered data on China's nature reserves nationwide and submitted a report thick with suggestions in 1993 that also included a section specific to the protection of the giant panda.[8] The work for this report was conducted between 1989 and 1992 as part of a cooperative project involving the Ministry of Forestry and WWF.[9] According to this data, Pingwu was the county with the country's largest panda population.[10] By 1996, the WWF had just gained official presence in China, established a China office, and hired Dr. Lü Zhi as its panda program officer upon her return from a post-doc in the United States. Lü Zhi, who had worked with Professor Pan Wenshi on her doctoral research on wild giant pandas, quickly became recognized as one of China's top specialists on the giant panda. After visiting Pingwu County and a few of the reserves within it, Lü Zhi decided to conduct a special project in the Wanglang area. Many have credited Lü Zhi with spearheading the transformation of Wanglang from a somewhat neglected reserve into a model reserve.

Since the 1976 bamboo flowering and die-off episode in the Min Mountains in and around Wanglang, bamboo had rejuvenated within the reserve to a level that could sustain a larger panda population. Logging within the area of the reserve had, for the most part, ceased in the 1960s when the reserve was created. Logging outside of the reserve, however, was pervasive and destroying adjoining panda habitat.[11] Lü Zhi was very concerned about the degradation logging brought to the area and especially to the panda habitat. At the same time, she advocated for the basic principle that conservation efforts will only succeed when the economic well-being of local people is met. WWF representatives thought ecotourism could serve as a means for the local Baima people to meet their economic needs without engaging in the habitat destruction associated with logging. Lü Zhi began pursuing the idea of transforming the nature reserve into both a scientific research base and an ecotourism site.

In 1997, Chen Youping became the new director of the Wanglang reserve. Lü Zhi found him to be "a man of initiative" who was receptive to cooperation with the WWF.[12] This was an accurate assessment of the energetic new director whose guiding principle for his new nature reserve management job was, "Make the reserve and its management better every year."[13] Directors of the Wanglang reserve traditionally were Party appointees, and Chen Youping was no exception. He had served both in the police force and the navy, which meant that he could enforce poaching regulations. Given that directors typically had no background in forestry, there was no guarantee that they would consider the scientific or environmental consequences of policy decisions. Chen Youping, however, was quick to recognize the value of cooperating with an international organization like the WWF. In addition to being able to offer nature protection expertise, the WWF provided funding, something the reserve was in dire need of. From his perspective, the greatest hindrance to improving management and transforming the reserve was accessibility. Neither patrolling for poachers nor tourism would succeed without better roads to the reserve, so he aggressively pursued the task of building a paved road. This goal coincided with the long-expressed desire of many members of the Baima community, who had hoped that the creation of the reserve would bring about the building of a decent road to their community.[14] A paved road, guest houses, and a lodge, pictured in Figure 7.1, in the reserve were all completed in 2000.

Chen Youping also benefited from a worthy deputy director, Jiang Shiwei, who was already active in the reserve office when he arrived. One of the few staff who had formal forestry education, Jiang Shiwei had worked at the reserve long before it became a WWF project. Staff had described it as undeveloped and lacking in economic resources. Jiang noted that little could be done without adequate funding. The government took little interest in it, and neither the county government nor the county forestry bureau contributed much money to it. Once the WWF put a solid development plan in place, along with some economic support and scientific expertise, "things began to change—the government took notice and began to offer much more financial support."[15]

In Lü Zhi's recollections of her first visits to Pingwu as a WWF Species Program Officer in 1996, logging weighed heavily on her mind. "Sixty logging trucks left the area every day," she noted, "timber was moving at the highest rate seen in the past ten years." Before the WWF could implement anything, Lü Zhi insisted that they needed to do more analysis and find out who would be affected by any changes in logging policy. "This kind of decision making," she continued, "needed to be scientifically based and participatory." By participatory, she meant that it required input from the local people who would be affected. Lü Zhi asserted that without the cooperation of local communities, new conservation policies would

FIGURE 7.1 This lodge and associated guest houses were completed in 2000 in order to accommodate the introduction of ecotourism and other programming in the Wanglang Nature Reserve. Courtesy of Wanglang National Nature Reserve, Pingwu County, PRC.

be useless. A community-focused subsidiary program of the WWF called Integrated Community Development Programme (ICDP) organized community surveys of the local Baima villages under the leadership of a young and energetic worker for the WWF, Li Shengzhi. Their results confirmed a strong dependence on the timber industry.[16]

Logging and panda preservation have come into direct conflict repeatedly because the forests that are attractive to loggers make prime panda habitat. Large coniferous forests foster a good undergrowth of bamboo. Edible bamboo requires an upper story of large trees in order to grow to the size and density that pandas favor.[17] Pandas also use large hollowed-out trees as birthing dens. These factors make old-growth forests not only the ideal habitat for giant pandas, but also a necessary environment for their continued existence.[18] Lü Zhi's Ph.D. advisor, Pan Wenshi, successfully campaigned against the logging operation in the panda habitat of Shaanxi Province and even managed to shut down the logging in that area.[19]

Lü Zhi saw a similarly dire situation in the Pingwu Mountains and sought to minimize the local dependence on logging by introducing ecotourism as an alternative means to generate a livelihood for the local people. As the ICDP coordinator, Li Shengzhi began refining details for the ecotourism plan. The WWF and ICDP were working to try to convince the local Baima that logging was destroying their environment and that there were other ways to make money. Lü Zhi, Li Shengzhi, and other members of the WWF team promoted tourism as an industry that could be both lucrative and environmentally friendly. Often this was a tough sell to people with little or no exposure to the tourist industry and whose family's daily sustenance was at stake. For the remote mountainous county of Pingwu, tourism was a slow-growth industry.[20] While taking an interest in developing some tourist sites and transportation, the reserve staff and WWF found the local population much more comfortable with the well-established timber industry.

Just as Lü Zhi and the WWF were struggling to foster enthusiasm for tourism among the local Baima people, the PRC national government handed down a decision that forced the shift. In 1998, the national government instigated a widespread logging ban across the areas that fed into the Yangzi River. In response to catastrophic floods along the Yangtze River that killed more than 3,000 people and created economic losses of over 300 billion yuan (about US $37.5 billion).[21] The ban suddenly prohibited all logging in the mountains of panda country. Consequently, finding alternatives to the timber industry was no longer a matter of choice. "The logging ban simultaneously imposed a great deal of pressure and created a tremendous opportunity," said Lü Zhi. With the exception of some subsidy, the government did not facilitate a transition to other income-generating occupations. The WWF had to act fast in the wake of the ban to facilitate its plan.[22]

Lü Zhi rebuilt connections between the nature reserve and the scientific community. She introduced the reserve to her Peking University colleagues, Wang Dajun and Wang Hao, who subsequently conducted extensive studies of panda habitat in the reserve. This not only resituated science at the center of reserve activities, but also gave the scientific community a voice in the continued development of the reserve. Wang Dajun and Wang Hao played a key role in its transformation into an ecotourism destination while they simultaneously engaged in deepening the reserve's identity as a scientific research base. Their field knowledge, commitment to both their study and the reserve, and their generous contributions of time and research resources all contributed to collegial relations between the reserve staff and other scientific experts they introduced to Wanglang.

Wang Dajun and Wang Hao also earned the trust of the reserve directors, who sought advice in making sure that management decisions were not in conflict with the ultimate goal of protecting the giant panda. Such an arrangement resolved many of the problems that critics saw with the tradition of appointing directors without background in forestry or biological science. This arrangement, however, was circumstantial, not structural, and temporary, which limited Wanglang's ability to truly serve as a model. The WWF thus was fortunate to have a staff on the ground that was eager to work toward making fundamental changes to the reserve itself. Because the concept of bringing ecotourism to Wanglang was initiated by scientists with ecological concerns, the development of ecotourism there was always in tandem with improving scientific understanding of the reserve, the pandas, and their habitat.

Coloring Ecotourism with an Ethnic Touch

In order to minimize the impact of ecotourism, the original plan was to have small groups of foreign ecotourists visit the reserve. These ecotourists would see the environment where pandas lived, learn about the animal and its plight, and take guided walks. The guide would point out evidence of panda presence in the area, such as scratches on trees, bamboo stands on which pandas had feasted, and panda scat. Because wild panda sightings are so rare, tourists were told ahead of time that it would be highly unlikely that they would see a live panda in the wild and that their trip was to experience a panda habitat and learn about the animals. To make sure that tourists did get to see live pandas (though in captivity) tours would stop at the Chengdu Breeding Center as part of the tour.

The next step was to engage the Baima people to participate in this enterprise. Initially this was not easy as tourism of any kind was foreign to this community. With the logging ban in place, however, convincing the local Baima to participate in this new tourism plan was much easier than before. The suddenness of the logging ban put many Baima in the unfortunate position of being in desperate need of more income but with few means of achieving it. In order to figure out exactly how the local Baima could contribute to the development of tourism, Li Shengzhi of ICDP and Wanglang's deputy director Jiang Shiwei became heavily involved in more extensive and detailed economic and sociological surveys of the Baima communities. The surveys offered promising results because the Baima people, as it turned out, possess a great deal of tourist allure. It is still common to see Baima people engaged in their everyday activities wearing their spectacular traditional dress, composed of brilliantly colored, woven belts displaying beautiful geometric designs tied around the waist of an ankle-length dress. Even more colorful than their belts are the women's blouses. The sleeves made with strips of brightly patterned cloth sewn together give the effect of a bold rainbow from their shoulders to their wrists. Men's style of dress is usually light in color, with a distinctive collar, and worn over common western-style clothes. Men and women both wear extraordinary hats of goat hide, with shallow crowns and medium-width brims scalloped around the edges. The crown is decorated with red and blue stripes configured from wool yarn wrapped around the crown several times to create the effect of a wider hat band. Tucked in these yarn bands are two very long rooster feathers tilted backwards that bounce playfully with the wearer's movements. Figure 7.2 shows women and girls from the Baima community in their ethnic dress.

When people drive through Baima areas they see workers in distinctive dress, out in the fields herding sheep and cattle, people tending chickens, and in the market place. Tourists, if observant, notice that this is genuine

FIGURE 7.2 These women and girls from a Baima community are wearing their traditional dress and distinctive hats. Photograph courtesy of Wanglang National Nature Reserve, Pingwu County, PRC.

everyday attire, not simply ethnic dress donned for the benefit of assumed tourist expectations. The Baima villages are conveniently located en route to the Wanglang reserve, which allow tourists an opportunity to visit the Baima community either on the way to the panda reserve or on their return. With the continually expanding market reforms, Baima could participate by hosting tourists, offering traditional food, and entertaining them with ethnic song, drink, and dance. The Baima could generate additional income by selling handicrafts, such as their brightly patterned belts and blouses.

Without concerted effort to create a route to bring tourists to the Baima community, few would otherwise encounter this historically isolated people, "whose language, customs and culture are different from any other ethnic group."[23] The experience of interacting with the Baima community is indeed unique. Because the tourists were unlikely to spot a live panda in the wild and because the landscape of the Wanglang reserve, though pretty, is not particularly distinct from Western mountain ranges, the Baima people, their folk songs, food, dance, and distinctive ethnic dress constituted the sole aspect of this tourist itinerary that genuinely offered visitors something that they could not see anywhere else in China or, for that matter, the world. The concept of doing ecotourism at Wanglang originated as a means to offer the local Baima people an alternative livelihood, but the success of this ecotourism program actually depended on the Baima.

During the mid-1990s, when the WWF initiated this project, ecotourism exploded as a new industry across the globe.[24] A combination of increased attention to the impact of "regular" tourism on local economies and the environment and an expansion of the environmental movement during the 1990s made it easy to find support for this pilot program. Such a fast-growing industry was difficult to regulate, but the WWF planned to "do it right" and use it to resolve the conflict between panda conservation needs and the needs of the local people.

Ecotourism was still quite foreign to China when the Wanglang ecotourism project began in 1996, but the reserve staff, local forestry bureau, and the Baima themselves were all being educated on its principles. The WWF was adamant about ensuring that ecotourism in Wanglang would be done in a responsible manner. They designed it to include environmental education as part of the tour, limit the environmental impact of visitors, and designate some profit from this venture to the Baima and some to maintenance and promotion of nature protection projects in

Wanglang. The Wanglang reserve staff and the WWF partnered with a California-based ecotourism company, KarmaQuest, whose mission was "helping to conserve the natural environment, endangered wildlife, and our world's rich cultural heritage by supporting conservation and sustainable tourism."[25] The company was chosen in part because it had been leading expeditions in Asia since the 1980s and had won awards for their tours.[26] In order for Wanglang to add to its legitimacy as an ecotourism site, Wanglang also invited an Australia-based ecotourism certification company, the Nature and Ecotourism Accreditation Program (NEAP), to assess the reserve and its tourist operations. This company evaluated the accommodations, energy and water usage, trash disposal, and tourist activities to make sure that the site and program could be characterized as ecotourism.[27] The NEAP representatives accompanied the first two groups of ecotourists in 2001 and awarded Wanglang an ecotourism certification that same year. With the combination of this ecotourism expertise, the novelty of being able to trek through panda country, and the allure of engaging with an exotic ethnic group, the prospects for fruitful growth in ecotourism to Wanglang were promising.

The interactions among the WWF, the Wanglang reserve, and the Baima people inspired and required adjustments. As neither Han Chinese nor Westerners ordinarily wear eight-inch wide woven belts, WWF representatives advised the Baima women to make more practical souvenirs, such as handbags. It also became apparent that this type of tourist industry favored women's skills. The ICDP, therefore, began investigating ways for the men to more actively participate. Recently unemployed truck drivers from the logging industry needed new ways of contributing economically. Beekeeping was traditionally a male craft that could also generate Baima honey for sale.[28] See traditional Baima beehives in Figure 7.3. Although at first the need for the inclusion of men was an urgent problem, as the tourist industry expanded, work opportunities for men did too. Men found work building lodges and increasingly were integral to ethnic dance and other ethnic entertainment.

Some of the utopian visions of this Deng-style cooperative-for-profit faltered with the realization that not everyone benefitted under the market model. Discontent brewed among a group of women who pooled resources to make handicrafts and left one woman in charge of selling them. Not everyone's handicrafts sold equally, and the group was dissatisfied with the resulting distribution of income.[29] The WWF website also noted that tourist lodge after tourist lodge was being built, raising anxiety among

FIGURE 7.3 Beekeeping and honey production have long been a component of Baima male culture. With the introduction of ecotourism, Baima honey became a local specialty for sale. Photograph by E. Elena Songster, Pingwu County, Sichuan, PRC, 2001.

some that there would be more beds than guests and that competition would make the venture unprofitable for all.[30] Poor Baima households did not have resources to build lodges or surplus to make handicrafts for sale, so could only watch as the new prosperity from ecotourism spread around them. Yet tourism continued to grow.

In 2001, five years after the idea was first put forth, ecotourism was well under way in Wanglang. In 2002 the Wanglang Nature Reserve succeeded in ascending from the status of a provincial reserve to that of a national reserve, a categorization more likely to attract more tourism. The status change demonstrated that the nature reserve's efforts to improve its facilities and management, partner with a strong international organization such as the WWF, integrate local communities, and upgrade environmental technology all met with the approval of the State Forestry Administration. The main motivation for applying for national status was to be able to tap into national funds. The status in some ways rewarded the reserve for pursuing creative ways to generate their own income; by no means did the government encourage or want Wanglang to change course in this regard. In fact, the State Forestry Administration saw Wanglang as a model and promoted it as an example for other reserves to become more creative fund raisers. Highlighting the efforts of this panda reserve also accomplished the goal of demonstrating that the government was putting greater emphasis on the environment. This was part of a shift that some scholars refer to as the establishment of an "ecological state" in describing the governments' general shift in emphasis, both genuine and superficial, toward environmental policy and protection.[31]

By 2001, ecotourism projects had impacted about 8 percent of China's nature reserves. The Chinese government had invested relatively little in nature protection; even as late as 2008, it spent only about $50 per square kilometer as compared to $150 in other developing countries. China's dollar-to-square-kilometer ratio was less than 1 percent of what was spent on the same area of protected ground in developed countries.[32] With such dismal government input, reserve staff increasingly saw ecotourism as necessary to supplement income.

The government responded positively to Wanglang's early successes. In 2005, the Mianyang Prefectural Forestry Office wrote up a report on four ways the Wanglang and Tangjiahe nature reserves had demonstrated initiative and innovation worth emulating. The prefectural office noted,

> Building a robust management system is the foundation of a reserve's management structure; developing the economy in the local community is an effective means of solving the contradictions of the community. Maintaining high quality ecotourism as a goal is critical to the sustainability of ecotourism. Conducting cooperative international research is an important factor in competing for funds to advance the development of the reserve.[33]

In subsequent years, the Mianyang prefectural office continued to sing Wanglang's praises in environmental education development. The Sichuan Provincial Forestry Department reported on the reserve's cooperation with such groups as Conservation International. Even the national SFA posted a report on their website titled "Wanglang: The Quiet Blossom Deep in the Heart of the Min Mountains" that praised the many developments under way at the Wanglang reserve.[34] Although this report was written by the reserve's own deputy director, the SFA's posting indicated that it wanted the government units under its broad umbrella to learn about the activities and accomplishments of this small, remote reserve. The last line of this article boasted of Wanglang's many achievements and awards for particular programs, most of these awards which recognized positive steps toward self-sufficiency.

The broad expansion of tourism of all varieties during the twenty-first century remapped China's landscape by transforming conservation strategies. This expansion coincided with the Chinese public's increased ability and growing interest in venturing into China's wild areas for both ecotourism and ethnic tourism. On all fronts, Wanglang enjoyed a privileged position as a testing ground for new approaches to conservation. The success with which Wanglang pursued revenue-generating approaches to the mission of the reserve, namely panda preservation, forest protection, and scientific research, not only attracted the attention of government officials, but also the attention of others aspiring to create ecotourism sites.

A natural place for expanding the ecotourism model was at Wanglang's sister panda reserve, Wolong, where many of the challenges associated with ecotourism worldwide surfaced quite quickly. The fact that prospective merchants perceived the project in Wolong as quick money was made clear in the number of external parties that flocked to participate. Half of the hotels were operated by external investors, which diminished the success of the overall project from the perspectives of the local inhabitants and ran counter to the idea that ecotourism should be promoting local economies to help local people.[35] The paramount goal of preserving the environment seemed to be among the most challenging. For one, a human population lives within the Wolong reserve. See Figure 7.4 for an image of a typical dwelling and small farm within the Wolong Reserve. The people whose lives have had the greatest impact on giant panda habitat are those who live deep in the forest areas and harvest timber for fuel wood and other household needs. Finding a different form of income for these people would be crucial for preserving panda habitat. These local villagers, however, have not been helped by the development of ecotourism in the reserve because the tourist activity was set up too far away from their dwellings, so they had no choice but to continue farming and logging.[36] Despite the fact that Wolong more easily attracts tourists than Wanglang because of the famous breeding center on site and its much closer proximity to the provincial capital, these other factors make it more challenging for ecotourism to truly succeed there. Figure 7.5 shows two pandas at the Wolong Breeding Center. The development of ecotourism in both panda reserves indicates the pervasiveness of the basic

FIGURE 7.4 This is a typical dwelling of a household inside the Wolong National Nature Reserve boundaries. Photograph by J. Matthew Diffley, Wolong Panda Center, Wolong National Nature Reserve, Wenchuan County, Sichuan Province, PRC, February 2002.

conflict between local population needs and habitat depletion, which is by no means unique to panda country. Interest in expanding ecotourism also reflects contemporary China's rapid economic growth and its citizens' exploitation of entrepreneurial opportunities.

Nature reserves that focused on other species saw potential in the experiences at the Wanglang panda reserve. After examining conflicts between the sustenance needs of the poor local human population and the preservation needs of the white-headed leaf-eating langurs he was studying, renowned biologist Pan Wenshi came to believe that the best way to save the langur monkeys was to devise ways to aid local people economically. His approaches included educating the local community and trying to develop a source of fuel other than the trees upon which langurs are utterly dependent. After he witnessed the programs his former students initiated in Wanglang, Pan also sought support from local government to build accommodations to house ecotourists and create an ecotourism program in the langur reserve he helped create in southern Guangxi Province's Chongzuo County.[37]

FIGURE 7.5 The allure of viewing pandas like these two attracts visitors from around the world to the Wolong giant panda breeding center, thus making the initial introduction of ecotourism logistically easier in Wolong than in Wanglang. Photograph by J. Matthew Diffley, Wolong Panda Center, Wolong National Nature Reserve, Wenchuan County, Sichuan Province, PRC, February 2002.

A successful ecotourism destination does not necessarily depend on a top-tier endangered species. A Tibetan community in northwestern Yunnan Province, for instance, began pursuing ecotourism in order to protect their small village and its surrounding area. Also subject to the logging ban that affected communities around the Wanglang reserve, the Tibetan people in Jisha County, Yunnan, needed to find a new means of income in 1998. In 2000, two Yunnan scientists, Li Bo and Xie Hongyan, succeeded in securing funds from the Global Environment Facilities and the Asian Development Bank to transform Jisha, into an ecotourism site.[38] In order to learn how to build accommodations and structure the tourism, representatives from Jisha traveled to Wanglang and the Baima villages.[39] By 2003, Wanglang had earned a reputation as a model ecotourism site. In 2010, Wanglang was honored with the National Geographic Traveler award for sustainable tourism management.[40]

Although Wanglang had become a model to many, not everyone approved of the reserve's new direction. A former staff member who had been observing the changes over the years stated plainly, "I do not like what they've done to Wanglang. . . . They are only paying attention to money. What is going on now is not nature protection. They are all getting carried away, carried away with money, and forgetting what is important."[41] From this perspective, the income-generating activities were a distraction from the basic tasks of providing a protective environment for the giant panda and enabling basic research to take place in the reserve. At the time, this was a minority opinion; reserve staff initially did not heed such concerns because they were receiving affirmation through formal recognition and others who were trying to emulate their work. The Wanglang ecotourism project, however, was subject to quick and unexpected turns.

From Baima (In)Dependence (Back) to a Monitoring Model

Even though the Wanglang program for ecotourism depended on the Baima, the Baima quickly grew independent of the reserve. Once they gained experience with this new industry, ecotourism to the Baima community soon transformed into ethnic tourism, which was already widespread in China and was particularly popular in Southwest China where many ethnic minorities lived. The Baima people who embraced tourism grew so successful that they became self-sufficient. Dramatically accelerated economic growth in China's urban centers led to a transformation of both ecotourism and the Baima community. Within the first few years most tourists to the Baima villages increasingly were Han Chinese who stopped there without venturing to the panda reserve or learning about the environment. The once small village was completely remade with a grand gateway into the town that was crowned by a huge, three-dimensional rendering of a traditional Baima goat-hide hat, see Figure 7.6. The motor traffic to the Baima village even justified the placement of a freeway sign directing traffic to the "Baima Tibetan Ethnic Township."[42] In addition to this, one can also find billboards (See Figure 7.7) advertising the Baima community, something that would have been unheard of just a few years earlier. Involving the Baima in an environmentally friendly activity ironically resulted in the Baima once again finding themselves on the wrong side of nature protection and creating yet another conundrum for the goal of panda protection.

Wanglang became a victim of its own success. Tourism was designed to be a tool toward creating economic stability for the local Baima. Unprecedented economic success led the Baima to dramatically expand

FIGURE 7.6 This village gate in the shape of the Baima hat is indicative of the economic success of Baima tourism and the growing recognition of the distinctiveness of Baima dress. Photograph by E. Elena Songster, Pingwu County, Sichuan, PRC, 2013.

FIGURE 7.7 This billboard highlights the association between the Baima people and the giant panda introduced with the original configuration of ecotourism by the Wanglang Nature Reserve. Photograph by E. Elena Songster, Pingwu County, Sichuan, PRC, 2013.

their livestock holdings, which in turn created new challenges for the reserve. Because many of the factors that led to the reserve's ability to generate income and elicit praise from officials were circumstantial and not structural, there has been no way to assure continued devotion to conservation principles or even maintenance of past achievements in spite of the best efforts of the reserve staff and administration. The established power structures beyond the reserve put the reserve at the liberty of China's layered governmental hierarchy, which encompasses a much wider and potentially conflicting set of concerns than those of the reserve itself. Upper levels of government demonstrated that ultimately balancing relations with local people outweighed reserve staff concerns about habitat protection priorities. The government has allowed the Baima people to pasture their expanded livestock holdings in the Wanglang reserve and the reserve staff, which consists of lower level bureaucrats, has been unable to effectively object.[43] As a result, China's oldest panda reserve now hosts growing numbers of Baima livestock, specifically horses as seen in Figure 7.8, and a cow-yak hybrid, shown in Figure 7.9, which have affected

FIGURE 7.8 These horses living and grazing within the Wanglang National Nature Reserve belong to people in the neighboring Baima community and are competing with giant pandas for food. Photograph by E. Elena Songster, Pingwu County, Sichuan, PRC, 2013.

FIGURE 7.9 These are cow-yak hybrid cattle living within the Wanglang National Nature Reserve. Although they are not perceived as having as much impact as the horses, they also are depleting and degrading panda habitat. Photograph by E. Elena Songster, Pingwu County, Sichuan, PRC, 2013.

panda habitat and the tourism environment. Recently a Chinese doctoral researcher from Duke University, Binbin Li, who was inspired as an undergraduate by the work of Lü Zhi and other conservation biologists, decided to do her field work in Wanglang. Binbin Li and her colleagues discovered that both horses and this specialized hybrid cattle directly compete with the giant panda for bamboo and the presence of these two forms of livestock have measurably decreased the panda's range within the reserve.[44] The reserve has gained renewed appreciation for the value in focusing on the original mission of the reserve, to protect and preserve pandas through scientific research. Moreover, it increasingly appreciates the inherent conflict between preservation work and income generation. This transformation raises questions about the real legacy of the Wanglang Reserve. In spite of the intensive energy and focus that the WWF and reserve staff devoted to the ecotourism project, from the perspective of Lü Zhi and other scientists involved in the project, ecotourism was but one of a wide array of tools that could theoretically improve the protection of China's wild population of giant pandas. In their view the real focus for the Wanglang reserve has always been improving methodologies of scientific study and better equipping the staff to monitor the reserve and its pandas. Cotemporaneous to ecotourism building efforts, the reserve had been continually engaged in scientific work and actively making itself a model in scientific conservation strategies.

While ecotourism grew, Dr. Wang Hao and others had been designing and expanding a wildlife monitoring system in the Wanglang reserve, the first need identified during the 1990s surveys and planning. Proud of Wanglang's historic role in providing pandas for panda diplomacy and participating in panda rescue efforts during the 1970s, the reserve staff wanted to continue to contribute to its legacy as a panda protection reserve. During the 1990s, the reserve staff had difficulty locating the giant pandas in their reserve and recognized that they did not have the training or expertise to find the famously elusive animals. As one of the visiting experts noted, "It is difficult to manage the protection of an animal without knowing where the target species is."[45]

Beginning in 1998, Lü Zhi and her fellow panda researchers from Peking University designed and conducted a panda survey protocol in Pingwu County, which became the foundation for a new panda monitoring program, subsequent research in Pingwu, and a pilot for the next national-level panda census.[46] The first monitoring program was established in Wanglang and became a base for training staff from many other reserves.[47] The basic premise of monitoring was to be able to track the presence, activities, and threats to pandas and other wildlife in order to establish and adapt preservation practices and policies. The concept of monitoring was not completely new to China, as Dr. John Mackinnon had introduced wildlife monitoring with the State Forestry Administration to China several years prior to Lü Zhi's first visit to Wanglang.[48] Some equipment was purchased, but not enough people were trained in using it to enable the program to take root. This experience made clear that monitoring had to be done in a way that was practical and easy for staff without scientific background to implement. Wanglang was the first reserve to use a digital Geographic Information System (GIS) that could process data from multiple sources and was much more flexible and useable than previous paper-recorded data. Mackinnon continued to train and work with staff at multiple panda reserves to expand this project, creating a rich data set that included monitoring routes, panda sightings, feces locations, and other indications of panda presence, such as chewed bamboo and scratched trunks. All of this data could be input with map coordinates. Scientists and trained Wanglang staff then began holding annual trainings to expand the functionality of the project and brought in staff from other reserves. In three or four years, several reserves began collecting and mapping data to expand the database. In turn, Wanglang quickly gained a reputation among its peers as an advanced reserve due to its own concerted efforts to define itself through scientific advancement.[49]

The monitoring system database was envisioned as a collaborative system that could both benefit from and facilitate ongoing and future research. Anyone conducting research in the reserve is invited to use and contribute to the database, and approximately one-third do. Although the mapping and monitoring program initially focused on giant pandas, the system was expanded around 2005 to include salamanders, small mammals, plants, and insects.[50] Everything came to a grinding halt in 2008, however, when the massive 7.9 magnitude Wenchuan Earthquake hit Sichuan province and killed more than 80,000 people.[51] All attention was redirected to the crisis, victims, survivors, and rebuilding. Earthquake-related construction lasted until 2013, after which there was renewed interest in resurrecting the nature reserve wildlife monitoring program in Wanglang and other Sichuan reserves.[52]

The twenty-first century evolution of nature reserve management in China involved multiple simultaneous metamorphoses. The original impetus for introducing ecotourism was to create a sustainable form of income generation for communities that shared habitat with pandas. At that time, no one expected the domestic economy and domestic tourism to skyrocket the way it did. Ecotourism was originally envisioned as a small-scale industry that hopefully would provide enough income to make local timber trade unnecessary and that required NGO funding because government funding was so limited. The combination of China's astounding economic growth, the various grants the reserve won, and government praise for the reserve's successful self-funding efforts all indicated that the nation's privatizing trend was moving into the hinterlands and to national reserves. Even though Wanglang had achieved national-level status as a reserve in 2002 and was able to tap into government funds that were previously inaccessible, it continued to pursue ecotourism and other income sources. When the colossal growth of domestic tourism consumed the carefully planned ecotourism framework and enveloped the Baima community with enthusiastic but not ecologically minded tourists, it made wildlife monitoring even more pertinent. After seemingly moving away from heavy and direct involvement in nature reserves, the government after 2010 began investing significantly more into the reserves and their programs.

Launching Forward from Its Legacy

Wanglang's role as a scientific research base was reinvigorated by the involvement of the WWF during the 1990s. The building of facilities and infrastructure for ecotourism only made Wanglang a more attractive place to conduct research. The integration of purposes was evident on the reserve's website in 2013. With bold red characters across the masthead of the home page, the Wanglang reserve invited visitors to conduct scientific research there: "Explore Nature's Ancient Mysteries, Discover Ecotourism in Wanglang, Construct a Cooperative Scientific Research Base."[53] The change from serving government-appointed scientists to serving anyone who browses the Internet demonstrates that hosting scientific researchers in Wanglang has been particularly good for the reserve in either generating revenue, prestige, or both. This advertising is far more public than typical of most research bases and reflects the current direction of the scientific community and scientific research. What constitutes a good research site for scientists is increasingly similar to that of a tourist site. Interested scientists can learn from specialized websites about accommodations at various research bases; the availability of GIS information; lists of local species; climate data; numbers of local staff members; and labs, potable water, hot water, telephone, internet access, and laundry service, in addition to more specific scientific needs.[54] Previously knowledge about the facilities at a remote scientific research site for wildlife fieldwork would have been limited to networks of specialists.

Another feature of Wanglang that it shares with the Wolong panda reserve is the legacy of scientific research conducted there. Existing scientific data is particularly useful for comparative studies, but also for laying a foundation on which to build related data. For instance, after the seminal giant panda survey was conducted in Wanglang from 1967 to 1969, the reserve hosted, among others, a Smithsonian study in the mid-1980s, a bamboo study in the mid-1990s, and more recently a wild panda survey that challenged the findings of the PRC government's Third National Panda Population Survey.[55] Published studies also make it easier to apply for grant money for further scientific research and conservation projects, as seen in the wide variety of Wanglang projects funded by the Critical Ecosystem Partner Fund (CEPF). These projects include panda corridor construction, environmental education through signage and boardwalk construction, anti-poaching studies, and comparative ecotourism development in other reserves around China. Since the first international research project on wild giant pandas began, the projects have been connected to the Chinese scientific and bureaucratic communities, as well as the landscape in which the research has been conducted. The reserve also has pursued a host of other scientific projects in which scientists from many Chinese scientific institutions have been involved.

The reserve is beginning to see its various roles converge. Scientists are studying the ecological impact of policy decisions. Dr. Binbin Li's research on the ecological impact of minimally managed Baima livestock, on the habitat of the giant panda is an example of doing scientific research on the reserve as well as in the reserve.

A Twenty-First Century Model

With this renewed focus on scientific study and preservation, nature enthusiasts now will revel in the knowledge that the Wanglang Reserve will continue to conserve wildflowers, wild mushrooms, and wild pandas. After transforming the underlying approach and purpose of nature reserves and wildlife management in China, the reserve returned its emphasis to science in recent years. The once isolated skepticism articulated by the former staff member, that the reserve should focus on protection and not get distracted by income-generating activities—even when they seem to be serving the needs of protection—has become a more commonly held belief. The reserve thus began to step back from ecotourism and instead pursued other forms of education as part of a network of conservation strategies, including educational outreach, workshops for staff at other nature reserves, and field trips to educate youth about pandas and nature. In so doing, the reserve continued to pioneer and model new services that a nature reserve could provide.

Scientific research in the reserve was originally justified in 1963 as "important service to the nation."[56] During the twenty-first century, the importance of the research conducted in Wanglang goes beyond the nation to the ecology of the region and conservation biology writ large. Modeling wildlife monitoring might not sound as exciting as pioneering ecotourism adventures, but it better reflects the true purpose of the reserve and produces an ever expanding collection of data for future researchers. By channeling more energy and resources toward the charter focus of the reserve, Wanglang staff are simultaneously contributing to a recent trend of linking the giant panda to environmental protection. This association has long been present at some level because of the long-held, if precarious, status of the giant panda as an endangered, and then vulnerable, species and its role as the symbol of the Worldwide Fund for Nature. As environmental issues command attention from ever widening numbers of groups within China and around the globe, the Chinese government increasingly recognizes that highlighting the association of the panda with environmental protection is a useful new form of diplomacy.

8

Olympic Pandas, Trojan Pandas, and the Science behind Soft Diplomacy

ATLANTA-BORN GIANT PANDA Mei Lan stepped onto a button on a platform in his new enclosure in the Chengdu Research Base, in Sichuan, China, and shut off the lights. More than 130 countries and hundreds of millions of people followed. With this act, Mei Lan launched Earth Hour 2010, an annual symbolic event of turning off lights for an hour to demonstrate a commitment to saving energy and protecting the environment.[1] The panda was serving as the World Wildlife Fund's official Earth Hour Global Ambassador, an honor he shared with esteemed dignitaries and celebrities such as African Archbishop Desmond Tutu, New Zealand Prime Minister Helen Clark, UK tennis star Andy Murray, and actress and Honest Company co-founder Jessica Alba.[2] Mei Lan's performance again reinforced the role of the giant panda as a representative of China to the world. What was striking in this particular display was China's participation in explicitly invoking the giant panda's symbolic association with environmental protection.

China's recent environmental record has been mixed at best, with most headlines focused on infamous dam building and catastrophic pollution. In the realm of panda protection in particular, China came under attack during the 1980s for favoring profit over the well-being of this precious species. Complaints included the continued presence of pandas in Chinese circus acts and the lax enforcement of protection laws restricting activity in the panda reserves. Criticisms from such wildlife protection organizations as WWF, the International Union for Conservation of Nature (IUCN), and the American Association of Zoological Parks and Aquariums (AAZPA) began to crescendo during the late 1980s with the expansion of a new form of panda diplomacy: short-term loans.

China began to clean up its environmental image in the mid-1990s with the introduction of a new form of panda loan—one that was based on scientific research exchange and designed to respond to criticisms of the previous decade. Panda diplomacy remained an accurate description of these exchanges, even as the terms under which pandas traversed the globe continued to evolve from gift, to short-term loan, to scientific exchange. Just as the new scientific panda exchanges were becoming established and gaining acceptance, China ruffled political feathers by offering a pair of pandas as a gift to Taiwan. Critics of this move called the gift an environmentally irresponsible use of pandas.

For China, the choice of one of its pandas as Earth Hour Global Ambassador was both an opportunity to begin to publicly improve its environmental record and an indication it had already made strides toward this end. The selection of Atlanta-born panda Mei Lan, a product of this new, more environmentally responsible scientific-exchange model of panda diplomacy, was part of China's effort to transform the meaning of panda diplomacy. No longer did the panda simply serve as a bridge between China and the world, but it presented to the world a more environmentally oriented China. Achieving this symbolic integration, however, was a multi-decade process that has required constant maintenance.

Olympic Pandas

In light of the bamboo flowering crisis of 1983 and other concerns about the sustainability of the wild population, the Chinese government decided to cease offering pandas as diplomatic gifts. This is ironic because the short-term loans that essentially replaced state-gift pandas was far more criticized by those advocating for the well-being of the species. The original rationale for the concept of the short-term loan was that if the pandas were simply lent to other countries, China would not lose any more of its precious pandas and, theoretically, the pandas on loan could be rotated through the breeding programs in China between their tours abroad. The loans, however, were not designed for the well-being of the pandas but followed market demand. A person involved in the management of the giant pandas at the Beijing Zoo compared the two forms of panda exchanges in this way: "There was no disgrace in sending these twenty-four [original state-gift] giant pandas abroad because they helped pave the road toward friendship between China and other nations."[3] The subsequent loans, he noted, were much more about money than international goodwill.[4] The first short-term loan was inspired in part by the Olympic

Games in Los Angeles in 1984. The decision to send a pair of pandas to the Los Angeles Zoo to commemorate this event and take advantage of the expected swell in attendance to area attractions spurred two significant trends in panda exhibits: first, short-term loans; and second, the notion of the celebratory panda, that is, pandas used to augment the festive atmosphere of special events.

The first short-term loan was such a success for the Los Angeles Zoo, for China, and the public that other zoos, fairs, and municipalities worldwide began to clamor for their own panda exhibits—even if they only lasted a few months. Because of the "heavy turnout of spectators," Los Angeles Zoo officials extended the panda exhibit for two weeks before sending the giant pandas on to the next stop on their tour, San Francisco.[5] The exhibits were bringing incredible financial benefits for those on both ends of these deals. Each agreement brought in several hundred thousand US dollars to China and foreign zoo profits ranging from US $3–4 million. These numbers inspired other cities to hope for even greater profits.[6]

Biologist and panda expert George B. Schaller analyzed the manic expansion of panda loan deals around the world in an essay entitled "Rent-a-Panda."[7] The scramble to send giant pandas on as many short tours to disparate points on the globe as possible reflected the exaggerated entrepreneurial ethos both within and outside the PRC at the time. Three problems he points to in his discussion of these loans became central to subsequent efforts in designing more responsible loans. First, Schaller noted that the most important problem with the loans, which varied in length from three months to a year, was that they disrupted the delicate breeding cycle of the animals, causing undue stress to the individual pandas. Breeding captive pandas was not yet successful enough to assure a sustained captive population.[8] This had a direct impact on the wild population because pandas that were sent abroad were being replaced with more pandas from the wild, which further threatened the stability of the wild population. Indeed, there was a sharp upsurge in the extraction of wild pandas during the mid-1980s when short-term loans were expanding rapidly.[9] Ultimately, endangered species advocacy groups determined that the short-term loans were harmful to the giant pandas—both the species and the individual animals.

The second problem was that the pandas had been transformed into cash cows for their host zoos and fairs, as well as for the Chinese government. This, as Schaller points out, was in direct violation of regulations established by the Convention on International Trade in Endangered

Species (CITES), which specifically limited the import of endangered species to those which the "Scientific Authority of the State of import has advised that the import will be for purposes which are not detrimental to the survival of the species involved . . . [and] a Management Authority of the State of import is satisfied that the specimen is not to be used for primarily commercial purposes."[10] The Convention therefore asserted that the movement of endangered species between countries was to occur only where the involved parties could demonstrate that the exchange would generate clear benefits for the species, advance scientific understanding of the species, would not harm the individual pandas involved, and was not primarily for profit; the short-term loans violated all of these policies.[11]

The third problem, and one that perpetuated these loans, was the inability to enforce CITES regulations. These were continually violated despite the concerted efforts to make participating organizations and government agencies aware of their egregious actions. The incredible profitability of the crowd-pleasing pandas proved more powerful than any regulations, but the real problem in enforcement, according to Schaller, was the involvement of government officials, at various levels up to the president of the United States, in brokering panda deals.[12] While high-ranking government officials get involved in supporting various types of international exchanges indirectly related to the business of government, the pervasive involvement of government officials in panda exchanges (even to the present day) reflects the close historic connection of these animals to high-level diplomacy. So even when diplomatic state-gift pandas ended, the short-term panda loans were strongly colored by an association between pandas and political figureheads, whose loan agreements were usually considered accomplishments by their constituents.

Following the incredible success of the panda loan to Los Angeles for the Olympics, Calgary actively sought to arrange a panda loan for the 1988 Winter Games. The notion that giant pandas were coming to be seen as a mandatory feature for Olympic host cities frightened the organizations that were trying to eliminate the short-term loans. Calgary succeeded in bringing a pair of pandas to the city, but only in the face of tremendous controversy. Meanwhile other major event coordinators vied for pandas to enhance anniversaries and state fairs. In 1992 pandas were brought to the city of Columbus, Ohio, which was honoring the five-hundredth anniversary of Christopher Columbus's arrival in the western hemisphere. The struggle to end all such short-term loans reached its peak of intensity right at the time the permit for the city of Columbus was being issued.[13] In spite of protest, the giant pandas became part of the city's events.

The use of celebratory pandas was also popular within the PRC, where particular attention was paid to the symbolism of numbers. Although the Beijing zoo already had pandas, it was seen as necessary that China move an auspicious eight additional pandas to Beijing to celebrate the start of the Olympic Games on August 8, 2008 (8/8/08).[14] In 2009, six pandas were sent to Beijing to honor the sixtieth anniversary of the founding of the People's Republic of China.[15] Six more pandas were sent to Guangzhou in 2010 to bring luck to the sixteenth Asian Games. Not surprisingly, China sent ten giant pandas to Shanghai for the 2010 World Expo. Celebratory pandas have usually been loans, but there are a couple of recent examples of celebratory panda gifts. China gave a pair of giant pandas to Hong Kong in 2007 to celebrate the tenth anniversary of the handover of the island from Great Britain to the People's Republic of China.[16] Another panda pair was offered to Macau in 2009 to celebrate the tenth anniversary of Macau's return to the mainland and the establishment of the Macau Special Administrative Region.[17] Because the resistance to global short-term loans curbed the continuation of this practice outside China, most of these recent celebratory events have taken place within China.

Theoretically, the concerns raised by those opposed to the short-term loans abroad, which mostly revolved around optimizing the reproductive potential of captive pandas, would best be served by keeping all pandas within China. Ceasing all overseas loans, however, was never up for negotiation. The only way that critics saw to bring an end to international short-term loans was to force loans to be negotiated under very different terms. The new policies written up in the early 1990s permitted loans to the United States that were to be based on scientific research, environmental educational outreach, and wild panda protection—the policies already established by CITES.

Scientific Pandas

The most common form of present-day panda diplomacy is the long-term, scientific loan. The stipulations for such loans are the outcome of extended efforts by zoologists and policymakers to adhere to basic regulations of trade in endangered species and create loan conditions that are as favorable to the species and individual pandas as possible, while still making zoo loans viable. The loan terms follow the basic formula of a loan of a male and female pair to live in a foreign zoo or park. The loan must also include a component of scientific research, research exchange, and environmental and educational outreach programs. The price for such a loan typically is US $1 million per year paid by the host zoo to China with a ten-year commitment. The fee is deliberately high so that it is difficult for US zoos to make money on the giant panda exhibits. All of the American host zoos have extended or plan to extend their loans beyond the ten-year agreement, but some have had to do extra fundraising to achieve this. The funds are required to be spent by China for the benefit of the wild giant panda population and involve participation of host-zoo representatives. When a baby panda is born in the context of such a loan agreement, additional money is paid to China and the cub is required to be sent to China after a few years (typically three) because, like the parenting pandas, it too is the property of the People's Republic of China.

Each component of these agreements is a response to a specific scientific or environmental concern. For example, exhibiting pairs that can mate enables the possibility of expanding the captive population and captive panda gene pool. Requiring long-term loans enables a period of stability for the individual pandas in question, which is considered to be both more humane and more conducive to reproduction. Requiring that funds for the display of these animals be applied directly to the wild population is a reminder that the theoretical purpose of having pandas in captivity is to educate people about them and to work toward the preservation and betterment of the wild giant panda population. The panda loan to the San Diego Zoo in 1996 was the charter experiment for this new model. It has been replicated in three other US zoos (Atlanta, the National Zoo in Washington, DC, and Memphis) and now many other zoos across the globe.

For many years, short-term loans of the kind that endangered species advocates tried to eradicate continued both within the United States and around the globe. Short-term loans continued to the United States as late as 1992. During the late 1980s and early 1990s, giant pandas also went on short-term loans to Canada, various countries in Europe, Japan, Singapore, and Thailand.[18] The Chinese government, after much urging, formally ended these loans in 2007 and required that all future overseas loans follow the model of the long-term scientific loan. China also reiterated that no more pandas would be given as gifts to foreign nations.[19] By ending the short-term loans, China could portray itself as more environmentally responsible.

While these loans cannot be characterized as diplomatic gifts, which serve as a public display of friendship and magnanimousness, they promote diplomatic interactions and exchanges. Diplomacy enters into scientific loans at many levels, including the interaction that occurs in the basic formation and upkeep of these arrangements, as well as with the management of the funds and the training, teaching, and scientific exchanges among the scientists and animal-care staff of the participating country. The most basic form of diplomacy, of course, occurs with the simple viewing of these animals by the public in the host countries and on the internet via panda cams.

The research-exchange model appeased most critics of the earlier short-term loans because it alleviated concerns that the Chinese government, scientific community, and viewing public were participating in an exploitative form of panda consumption. This policy solution demonstrates the intensity of the bilateral commitment to panda exchange programs when it became clear that the short-term loans to the United States would be abruptly discontinued. The fact that endangered species advocates had to fight so hard to enforce CITES regulations also demonstrates the sheer power of the panda's popularity. Much like the short-term loans, these new scientific loans continued to involve politicians ranging from city mayors and congressional representatives to the heads of state.[20] Even with the new protocols, most politicians benefitted from arranging a successful panda exhibit. Some exceptions to this general rule do exist. In Canada in 2013, Prime Minister Stephen Harper fell under fire when he chose to travel to receive the panda pair in Toronto instead of meeting a group of First-Nation school children who completed a 1,600-kilometer hike and snow-shoe trek to the capital to protest the conditions of First-Nation education in Canada.[21] This is a rare example of pandas inciting ire instead of affection upon their arrival. The fact that the prime minister chose to greet the pandas in the face of such a situation demonstrates the relative rise of China's status in global geopolitics in the eyes of world leaders and the increasing complexity of panda politics.

Cross-Strait Pandas

No panda exhibit created as much of a political uproar as the proposition of sending giant pandas to Taiwan. On January 26, 2009, the first day of the Chinese Lunar New Year, the giant panda exhibit opened for viewing in Taipei, Taiwan. News stations broadcast Taiwan's newly elected president, Ma Ying-jeou, strolling through the new panda enclosure, hand-in-hand with orphans and Taiwan's low-income children, who were visibly charmed by the pandas on display and this remarkable opportunity to be the first to view them. These images belie how controversial these pandas had been over the previous three years. Such a celebratory display of China's most famous ambassadors, Tuan Tuan and Yuan Yuan, on the most important holiday was a brilliantly choreographed demonstration of cross-Strait politics between the PRC and Taiwan.

The initial offer of giant pandas as gifts to the people of Taiwan by the government of the People's Republic of China in May of 2005 immediately ignited long-standing tensions that have defined the relationship between these two places since 1949. From some perspectives, the offer was a benign gesture of generosity and friendship, an offer too good to turn down. Yet to those sensitive to protecting Taiwan's nuanced *de facto* independence, it was simply a "charm offensive" against the island.[22]

At the heart of the dispute was the ambiguous status of Taiwan's sovereignty. Taiwan was in a very peculiar diplomatic position—neither a recognized independent country nor a province governed by the PRC. The PRC had been broadly recognized as the legitimate government of China since the 1970s, and China asserts that Taiwan is a province of the mainland but does not actually have any direct power over it. Taiwan enjoys its *de facto* independence but is wary of asserting an explicit declaration to transform its status into *de jure* independence, largely because the PRC government has threatened to use force against the island if it does.

From 1949 through the 1980s, there was almost no interest on the island in reunification with the mainland. In fact, Taiwan's Kuomindang (KMT) government had instigated the "Three No's" policy as a general principle on which to base its interactions with the government of mainland China: no contact, no negotiation, and no compromise."[23] The economic situation was also very different, as Taiwan's economy was reported to be one of the leading economies in the world, boasting 13 percent growth during the mid-1980s, while the PRC was recovering from mass campaigns, famine, and the political turmoil of the Cultural Revolution.[24] The difference in standard of living dramatically reflected their separate histories. Talk of unification across the strait through the 1980s was almost always a unilateral sentiment initiated by the PRC government.

In 1987, the PRC government decided to push the issue and offered Taiwan a gift of a pair of giant pandas.[25] Even with a single party dominating Taiwan and little division on the island over the issue of reunification, the government found itself under political pressure to accept the gift offer. When first proposing the idea in 1986, Liu Caipei, a mainland citizen who had been born in Taiwan and had returned to the PRC, argued that it would be good to send giant pandas to Taiwan because "This action would meet an enthusiastic response from the 19 million people of Taiwan."[26] This proved to be true. In 1987, the Taiwan populace wanted the pandas, but the government did not want to accept them, mainly for fear that such a move might be interpreted as capitulating on the stance of sovereignty. Although the president in 1987, Chiang Ching-kuo, did not yet have to answer to the public, the popularly elected legislature was not immune from the political pressure of public opinion.

After extensive debate, the pandas were rejected in 1988. The delay and deliberation in responding to the offer were testaments to the degree of caution the KMT government felt compelled to engage in. Officials ultimately used the rationale of environmental stewardship to decline the gift. The Agricultural Commission "announced that the offer was being turned down for conservation reasons."[27] According to Schaller, it was "a decision acclaimed by environmentalists."[28] At that time there was plenty of speculation about whether or not Taiwan could properly care for the animals. Newspapers conjectured, "The Taipei City Zoo lacks proper equipment and personnel to breed the animals."[29] Such concerns had much broader implications than the technicalities associated with raising exotic, endangered species. Just as accepting the pandas carried possible implications about the sovereignty of Taiwan, success in caring for them offered Taiwan the opportunity to demonstrate itself worthy of international recognition for its scientific prowess. At the same time, accepting the pandas carried the risk of failing in this task, which would damage Taiwan's efforts to achieve international recognition. Members of the Legislative Yuan in Taiwan who debated this issue during the 1980s put it this way, "If Taiwan can raise pandas, it will improve the nation's international image"; however, if they failed, that "would not only harm the bears themselves, but the nation's international image as well."[30]

The Taiwan government, however, found that it had to reject the pandas repeatedly because the PRC, sensing it had softened the hearts of the people in Taiwan, renewed its offer in 1989 and again in 1990. Having already debated the issue at length, the KMT government brought the final offer to a quick close in 1990 with an assertive statement that "the rejection was not a political move."[31] With that, "the Republic of China's Council of Agriculture closed its file on the panda plan, dealing a final blow to the proposal to relocate the rare animals."[32] In the context of the late 1980s and early 1990s, Taiwan's decision to focus on conservation concerns enabled the government to both successfully dodge implications of compatriot status in its relationship to the PRC and elicit praise from the international community for being environmentally responsible.

As China's economy skyrocketed and the two sides of the Taiwan Strait grew increasingly interconnected economically, hope again rose for those who wanted to see pandas in Taiwan. In 1995, giant pandas were explicitly mentioned in reference to the warming relations between Taiwan and Beijing. The original champion of the idea a decade earlier, Liu Caipei, resurrected it as perfectly relevant to the new "period of negotiation."[33] An official panda offer was made again in 1997.[34] The PRC government clearly felt this was an effective approach to winning goodwill among Taiwan's population, if not manipulating the local politics of cross-strait relations in Taiwan from the outside. Renewed efforts to expand bilateral exchanges between the island and the mainland in 2000 inspired the mayor of Taipei, Ma Ying-jeou, to champion the cause of bringing pandas to Taipei.

Trojan Pandas?

When the mainland again offered a pair of pandas to Taiwan in 2005, it presented the Taiwan government with a conundrum. The first politically problematic aspect was related to the structural change in the election of Taiwan's president. China made the panda offer to the then chairman of the Nationalist Party (another name for the KMT), Lien Chan, during his historic visit to the mainland that year. The KMT did not represent the executive power in Taiwan at the time; the president of Taiwan, Chen Shui-pien, was a member of the Democratic Progressive Party (DPP). Members of the DPP historically have been more vociferous advocates of formal independence than those of the KMT, and Chen Shui-pien was considered to be a thorn in the side of the PRC government during his years in office for actions and statements that the PRC perceived as pushing in that direction. The fact that the panda gift offer was made to the leader of the opposition party was a slap in the face to President Chen Shui-pien, who was not contacted at all regarding the matter.

The second politically charged aspect of this offer was the fact that the pandas were not being offered as an international state gift (a discontinued practice), but rather as a gesture of goodwill to "compatriots" in Taiwan.[35] Such a gift carried with it the implication that Taiwan was not another country. In the recent past, gifts had only been given to Hong Kong and Macau, areas that formally had become part of China. As such, Chen's administration and many in Taiwan were concerned that by accepting the pandas as a gift, the Taiwan government would in essence be accepting China's claim over the island.

The third element that made this panda offer unpalatable for the DPP government was the naming of the pandas. Before Taiwan had made any decision about whether or not to accept the gift, the PRC government invited its citizens on the mainland to vote on names for the specific pandas that already had been chosen to be given to Taiwan. During the PRC's televised pageant on the eve of the Chinese Lunar New Year in 2006, over a hundred million participants chose the names Tuan Tuan and Yuan Yuan. When combined, the names form "Tuan Yuan," which means "reunion."[36] This name choice was an overt reference to the political status advocated by mainland China for Taiwan.

In spite of all of this political controversy, the idea of having pandas in Taiwan was incredibly popular on the island. Polls commonly showed that more than 50 percent of Taiwan residents favored accepting the pandas.[37] Some polls indicated a figure of 65–70 percent, while others cited more than 70 percent. At the same time annual polls on the issue of unification showed that between 13 and 14 percent were interested in future unification with China, while between 20 and 24 percent had interest in future formal independence. Desire to maintain cross-strait relations at status quo was at 60 percent.[38] Such high percentages of those in favor of bringing pandas to the island thus indicates that even people who did not support unification wanted giant pandas in Taiwan. Those who fell into this group either did not consider the giant panda to be, as some feared, a "unification tool," or their desire to have the pandas on the island outweighed their concern about the implications associated with their presence.

The president challenged the PRC's insistence that the giant pandas were simply goodwill offerings with the proposition that goodwill could be demonstrated more effectively. As he noted:

> The Chinese government on the one hand deploys missiles against Taiwan, but on the other hand wants to give us a pair of giant pandas in the hope of winning the friendship of the people of Taiwan. If China genuinely cares about the friendship of Taiwanese people, why can't they dismantle the missiles aimed at us and use their missile budget for panda conservation?[39]

Those opposed to accepting the pandas began calling the animals "Trojan horses" from the PRC government, echoing sentiments that surfaced in the United States in 1972 (see Figure 5.2).[40] Debates in Taiwan about the pandas split between those who felt strongly that China was seriously meddling in island politics and those who felt it ridiculous that the animals could pose a threat to Taiwan's sense of sovereignty. Meanwhile environmentalists and Taiwan independence supporters dressed in giant panda costumes marched with signs that read, "Refusing to be a unification tool."[41] Nonetheless, the idea of the pandas remained popular. Taiwan citizens and companies donated millions of NT (New Taiwan) dollars for the new giant panda enclosure at the Taipei Zoo that ultimately cost NT $310 million (more than US $9.6 million).[42] The cost of panda care became a topic of contention as well. Those opposed to accepting them complained that too many resources were being thrown at the pandas. Those investing in the effort complained of wasted funds if the government refused to allow the pandas entry to Taiwan.

Because the pandas were so popular among Taiwan's citizens, President Chen Shui-pien had to be very careful in his response. He encouraged the Council of Agriculture to reject the pandas, and in a long personal essay, he framed his position around environmental stewardship. He asserted that pandas would not be happy in captivity on the island. Citing Pan Wenshi's research on wild pandas in mainland China, President Chen argued that pandas should stay in the wild and not be brought to zoos because, "Only by being able to wander freely in the wild can giant pandas smoothly perpetuate their existence."[43] Chen's reasoning was not entirely applicable to the situation at hand. The pandas chosen to come to Taiwan were captive born, not wild, nor did he completely reflect Dr. Pan's perspective. What was most significant about his essay, however, was that it focused completely on the environment and the welfare of the animals, with politics conspicuously absent. The Council of Agriculture ultimately declined to grant approval to the institutes that had applied to house the pandas. It did not address President Chen's concerns about the plight of captive and wild pandas, but rather focused on the inability of Taiwan facilities to fulfill the needs and responsibilities associated with housing these precious and endangered species. Neither institution, it asserted, had placed enough emphasis on environmental education and research:

> Most participants, including local conservation groups and experts, expressed their opposition to the import of giant pandas.

> The committee concluded that the import of protected species, according to the Wildlife Conservation Law should primarily be for the purposes of academic research and education. . . . The exhibitions and educational programs for strengthening wildlife conservation submitted as reasons to import the giant pandas by the applicants are far from concrete. . . . Neither applicant is comprehensively prepared in terms of the necessary breeding equipment and the training of medical and nursing staff.[44]

The Council of Agriculture concluded that it would not be good for the giant panda individuals or the species in general to accept the panda offer from mainland China under present conditions. After nearly a year of debate, extensive preparations, public protest, and widespread speculation, the panda gift offer was rejected at the end of March 2006.[45] The grounds for the rejection were very similar to those used in 1988, including parallels in the actual phrasing of the rejection.

Another conspicuous parallel between the two rejections was a strong resistance to addressing any political issues in the process. In a press conference explaining the decision, Lee Tao-sheng, deputy director of the Forestry Bureau and head of the committee in charge of reviewing the applications to host the giant pandas in Taiwan, made a point of explicitly depoliticizing the decision, saying, "There was no political influence on the committee's decision to bar the pandas from coming to Taiwan."[46] Of course, the first response from Beijing was that Taiwan had unduly politicized the giant pandas and the goodwill gift. Mainland newspapers editorialized, "There is absolutely no doubt that the rejection was made out of political considerations."[47] Completely unconvinced by the rationale offered by the Council of Agriculture, the mainland's main newspaper, the *People's Daily,* alleged: "The agricultural administration just implemented a political decision of Chen Shuibian by creating the technical reasons."[48] Accusations of politicizing the pandas were not made solely by the PRC government and its media mouthpieces. Leaders of the KMT opposition party also pointed to politics as the true impetus for the decision. The KMT's mainland affairs director, Chang Jung-kung, noted, "There is absolutely no doubt that the rejection was made out of political considerations."[49] In light of his historical advocacy for pandas in Taiwan, it is not surprising that KMT chairperson Ma Ying-jeou was critical of the decision of Chen Shui-pien's government. Chen Shui-pien responded by asserting that his government agencies had based their decision solely on conservation concerns, noting, "The pandas of course are cute, but in reality are living things, not toys, and should never be used as political tools. . . . Everyone must respect the experts' decision and should not make too many political associations."[50]

To be found guilty of politicizing the pandas was clearly political poison in Taiwan. No one, however, was immune from such accusations. Moreover, the one thing that all parties within Taiwan and across the strait agreed on was that tainting these wonderful animals with politics was fundamentally wrong. A special correspondent for the BBC who invoked the notion that pandas cannot be political went on to say, "Pandas do not belong to the CCP, nor do they belong to the KMT or the DPP, all people therefore should welcome them to Taiwan."[51] The great irony in all of this was that panda diplomacy has never been anything but political. While the PRC government has given giant pandas to allies such as the Soviet Union and North Korea as seemingly simple "goodwill gifts," it also has coupled panda gifts with two of the most famous political overtures of the twentieth century—normalization of its relations with the United States and Japan. In fact, panda gift offers were so successful historically in part because they accompanied such dramatic policy reversals by the governments involved. The PRC government also knows, however, that pandas have been, and remain, successful political tools only when the gesture of bestowing them on a given country is depoliticized, at least rhetorically. This process of depoliticizing the pandas can be seemingly successful because the animals themselves are so clearly incapable of engaging in politics directly. The PRC's panda offer to Taiwan not only was decidedly political, but it was also shrewdly staged at every step, with savvy timing and close attention to detail, which remained true in future dealings.

In January 2008, the Nationalist Party (KMT) won a landslide victory in legislative elections (which take place prior to the presidential election).[52] The party's presidential candidate, Ma Ying-jeou, advocated closer economic and cultural ties with mainland China and had also been a strong critic of the decision to reject the panda gift offer. During previous presidential elections in Taiwan, the PRC was accused of trying to influence the outcome in favor of KMT candidates with missile exercises. This year, the PRC decided to take a softer approach in its response to presidential elections on the island. On January 30, 2008, just six weeks before Taiwan's presidential election, a Taipei newspaper headline announced, "China Renews Offer of Pandas to Taiwan."[53] Taiwan's declining economy,

Chen Shui-pien's declining approval rating, and the promise of pandas all seemed to work together to usher in political change.

During a press conference on March 23, 2008, one day after his landslide victory, President-elect Ma Ying-jeou vowed to approve the acceptance of the panda gifts from China as soon as he was sworn into office.[54] Mainland China took note and in less than two weeks formally renewed their offer.[55] The PRC government demonstrated through the timing of these panda offers that it was highly aware of the political force of these animals and had a clear understanding of the efficacy of panda diplomacy. The PRC was rewarded for its efforts as Ma proved immediately to be a more PRC-friendly leader. After his election, he revised his party's former "Three No's Policy" into a much more moderate approach to mainland China: "no declaration of independence from China, no unification with China, and no use of force to resolve differences across the Strait."[56] The panda gift was in line with this broader shift in policy.

Ma Ying-jeou still had to be cautious about the way he framed the acceptance of the panda offer—ever aware of potential criticism that he had become too cozy with the mainland. In the same way that the rejection of earlier panda offers had been phrased apolitically, Ma's acceptance was couched in the language of environmental stewardship. The pandas were ultimately accepted in 2008 after the Council of Agriculture conducted another inspection of the zoo and deemed it qualified to care for the animals, because it had developed a plan to use them for environmental education and also provided a feasible proposal to contribute to research that would benefit the wild population. The environment and environmental stewardship thus became inseparably linked to the politics of panda diplomacy as the most politically safe means of facilitating the cross-strait panda gifts.

The new KMT government of Taiwan recognized that the best way to achieve peace on the panda issue was to try to further depoliticize the animals. The decision to debut them on Lunar New Year and to give low-income children and orphans a preview with the new president two days before the rest of Taiwan were clear indications that the politicians on the island of Taiwan had become savvy in the delicate details of panda diplomacy. The initial popularity of two of the most controversial figures in Chinese politics, Tuan Tuan and Yuan Yuan, was clear from the masses of excited spectators who lined up to see the pandas at five in the morning on the day the exhibit opened to the broader public. More than eighteen thousand people visited them on that first day. Moreover, the pandas produced an offspring in 2013, which, unlike the baby pandas produced during scientific loans, can stay in Taiwan and has received an honorary citizen card.[57] The extensive history of conflict and political tension that circulated around the pandas over the three years prior to their arrival did not dissipate immediately, however. When Chen Deming, president of the PRC's Association for Relations across the Taiwan Strait, visited Taipei in 2014, he made a special point of going to visit the cub, Yuan Zai. He was met by protestors at the zoo holding signs and shouting that China and Taiwan are two separate countries.[58]

The PRC, initially critical of Taiwan's tactics of employing environmental concerns in their efforts to navigate the politically charged waters of giving pandas as gifts, also has come to recognize that environmentalism is an integral part of the image of the panda and, as such, has added political value. As the PRC government emerged from its struggles and the challenges to its reputation for panda conservation during the late twentieth century, it forged its twenty-first century reputation on a conservation platform. Participating in such feel-good campaigns as Earth Hour, reflects the PRC's expanding use of the giant panda image to promote an ever widening array of other conservation projects.

Conclusion

WHAT DO THE blobfish, the Ohio lamprey, and the pygmy hog-sucking louse have in common? All are classified as "Endangered Ugly Things." This is not an official categorization, but the creator of the website dedicated to these animals invented it in an effort to "promote awareness of endangered species that wouldn't otherwise get noticed due to appearance or obscurity."[1] To this end, creator Nathan Yaussy has expanded his specialized endangered species list periodically. In 2010, he featured the species *Ailuropoda melanoleuca*, or the giant panda. One must certainly wonder why. That year, the black and white mammal could be characterized in many ways, including as endangered, but certainly not as ugly. To the contrary, it enjoys a seemingly perpetual consensus that it is an attractive animal. Moreover, it is not simply attractive, as one might characterize a peacock or an Arabian horse, but elicits affection in humans, irrespective of age or background. Even the panda's detractors do not contest its attractiveness but object to the fact that the panda commands so much attention solely *because* it is attractive or, more specifically, cute. One could argue that the Endangered Ugly Things website exists for all of the creatures that do not have panda appeal. The website has even sold T-shirts that read, "Forget the panda, save the Ohio lamprey." How did the panda become lumped in with a creature that sucks the blood of its victims through its circular, tooth-filled, jawless mouth?

Yaussy placed the giant panda on the Endangered Ugly Things list because he feared that the excessive attention to cute giant panda cubs and its iconic status had effectively removed it from the animal world. So, this "Endangered Ugly Things" exhibit attempted to "uglify" the panda by including links to videos of pandas attacking people and gluttonously munching on the leg of a fly-covered deer carcass. He featured these images in order to transform the oft anthropomorphized bear back into a typical wild animal. This message is important because ultimately the panda's value in nature, in science, and as a symbol of the People's Republic of China lies in the panda's primary identity as a wild animal. The desire to underscore the giant panda's existence as a wild species is in part a very material one that focuses on how to protect the animal's existence in its diminishing wilderness habitat but also encompasses the animal's symbolic significance. Diplomatic and scientific exchanges that surround captive panda research cannot even pretend to support the end goal of conservation without a wild population to serve and conserve. The continued persistence of the wild giant panda in the wilderness of China is necessary for it to maintain its present iconic function in China and beyond.

The symbolic importance of the giant panda's presence in nature stems from its surrogacy for its own home range and China's natural environment more generally. If the panda is China's, so too is the land from which it came. This creates a timeless congruency of nation and territory in the minds of the modern citizen, although the territorial boundary of China varied dramatically over the centuries and did not always include all of panda territory. The idea of the panda as a live, breathing animal existing exclusively within the present boundaries of the PRC reifies the myth of modern nationhood and helps bolster the legitimacy of the modern state and its present territorial boundary claims.

The giant panda became more broadly relevant to human society when it was deemed a scientifically interesting animal. Determining what kind of animal the panda is and where it fits into various taxonomic and evolutionary schematics dominated scientific discourse about the panda in the early decades of the twentieth century. Specialists from various fields of science have examined numerous aspects of the giant panda's animalistic traits, from its bone structure and diet to its reproductive practices and more recently its genome. Much of the thought, time, money, and energy invested in various forms of scientific research on the panda were all seeking insight into one of the most basic questions about the panda: what kind of animal is it?

This puzzle aroused scientific interest in the animal around the world. Being able to answer the many scientific questions about the panda became a point of national pride for Chinese scientists. The more unique the panda was among animals, the greater the national treasure it was. Even though the debate has carried on into the twenty-first century, early discussions on the topic were important to establishing the giant panda as a serious scientific subject prior to the founding of the PRC. This debate put the panda on the radar of the highly science-centric PRC government during the early 1950s.

Research on pandas has probed most aspects of the animal's being, from histological and nutritional studies to the much more high-profile work on captive breeding. Such persistent and wide-ranging scientific work on the giant panda by Chinese scientists over the course of PRC history reflects the tremendous value the Chinese state has placed on the panda. The ever-growing integration of cooperative expertise did not detract from China's desire to demonstrate particular expertise in panda science. The PRC currently boasts the most successful breeding record worldwide and is involved in such projects as panda reintroduction and mapping the panda genome. This wide spectrum of cooperative panda research reflects China's present position in the world: more open, more expansive, and more integrated with the global society, yet still highly competitive.

The giant panda was an important factor in helping to incorporate China into the international community both through science and diplomacy. The panda has been a particularly powerful political tool precisely because it is absolutely apolitical in its own right. For this reason, the government of the PRC is able to assert accurately that the pandas it has bestowed on foreign countries and loaned to foreign zoos have no political agenda. Again, this claim is accurate because the giant panda is an animal. As an animal (at least as far as we can discern), it is incapable of personal political perspectives. Because of the actual innocence of the animals themselves, the PRC government presents these gifts and loans and other types of panda diplomacy as apolitical goodwill gestures. The PRC government itself, however, cannot claim the same political innocence as these pandas. This form of soft diplomacy benefits the PRC government when any of the positive sentiments that people of the world have about pandas are then associated in some way with the nation of China.

Panda diplomacy works best when the people of the receiving nation or city come face to face with the live giant pandas. It is at this moment that the press and posturing that surround the gift or loan falls away for a moment. In the context of breeding-center and zoo exhibits, reminding people that the giant panda is but another animal is simultaneously unnecessary and unconvincing. While they sleep and pace in their enclosures like other animals, pandas also exhibit traits that have consistently inspired affection and invited anthropomorphism. The giant panda, like other bears, shares more basic physical traits with humans than do most mammals, such as having forward facing eyes, being omnivorous, being seemingly tailless, and even sitting in a very similar fashion. They look like they could embody human feelings and thoughts. Their coloring is almost clown-like with enlarged eye patches and a harlequin-like black and white pattern. Because they are virtual vegetarians and have relatively docile demeanors, pandas seem less threatening and less wild than other bears. Their faces and bodies are rounder and pudgier than most bears, making them look like teddy bears come to life. They looked so huggable to one observer at a zoo that he actually jumped into the enclosure and attempted to embrace one. This story demonstrates that the experience of being in the presence of the live animal ironically can undermine its identity as an animal. This paradox collapsed when the panda's space was genuinely invaded and it felt threatened by the man. In an effort to defend itself, the startled panda quickly displayed its animalistic behavior and seriously injured the man.[2] Most people do not test their desire to embrace pandas as though they were giant toy stuffed animals, and are thus able to preserve their anthropomorphized perception of them.

The generalized incapacity of humans to see the panda simply as an animal is magnified several-fold when observing panda cubs and mother-cub pairs. Tremendously popular zoo cameras that allow internet users around the world to watch panda mothers interacting with their tiny cubs expand this experience to all those who do not have an opportunity to see pandas live. Panda Cam followers erupted in public outcry when the service closed temporarily at the National Zoo during the government shutdown.[3] Zoo bloggers are emotive. One blogger on the San Diego Zoo site, noted, "This baby cub just melts me each time I see him! Bai Yun is such a good mom, I love to watch her tending to him. I am addicted to the panda cam and so thankful that you provide it for us to watch him grow."[4] People enjoy watching the interactions between the pandas because they reaffirm human expectations of a model mother-baby relationship. Another blogger proposed that in one video clip, "Bai Yun whispered in Mr. Woo (her cub)'s ear . . ."[5] These types of anthropomorphism are important because they exemplify the ways in which the giant panda is subject to the projection of human sentiments that have contributed to its iconic rise. As an animal, it is not verbal enough to reject any of these projections. People then are free to impose human feelings, thoughts, and actions onto the animal without much contestation. The fact that the anthropomorphism of pandas is convincing and that most of the sentiments toward pandas that result from it are positive have bolstered the panda's ability to successfully serve many different symbolic purposes over the decades. It is this characteristic of the giant panda that makes it so unique both in the animal kingdom and among other charismatic megafauna.

There are numerous cute animals—puppies, monkeys, and penguins, to name a few. Most of these animals do not exist exclusively in one country. They therefore cannot be as effective as a diplomatic tool, nationalist brand, or national symbol. Even the kangaroo, which is a country-specific animal that is used as a national symbol, is not as unique as the giant panda. The kangaroo is but one example of the many unusual marsupials that Australia has to offer, a few of which look quite similar to it. Kangaroos are not as beloved locally, as evidenced by the popularity of "roo bars" to protect cars from damage when hitting a kangaroo that is hopping across the road. It also is hunted for sport and the government issues special permits to cull its expansive population.[6] The kangaroo's compatriot, the koala, is arguably much cuter, but aside from appearing huggable, has little else to offer. Although the koala is not seen as a pest, the kangaroo has earned more symbolic recognition, appearing on the national crest and those of both Qantas and Australian Airlines. In part because of this domestic competition, the koala, unlike the panda, often is not the first animal one thinks of in association with its home country.

Most animals that represent countries are not cute, and for good reason. Nations generally want to evoke a sense of power through their symbols. The eagle, a favored symbol of many countries, including the United States, Mexico, and Germany, and the lion, representing India, Kenya, Singapore, and the British crown, are majestic, graceful, and intimidating. Such animals can be admired, but they do not elicit affection. The purpose of such national symbols is not to emanate friendliness and inspire adoration, but to exude strength and superiority. Images of such bygone beasts as the California grizzly and mythical creatures like China's imperial-era dragon, have been successful in holding strong symbolic significance because they invoke history or legend. The panda is not a creature with historical importance, mythological significance, literary symbolism, or even an important role in the economic or dietary lives of humans historically. Its lack of broadly recognized historical meaning bolstered its rise as an icon. It did not become an official symbol of the PRC in a debate over worthy animal emblems, but emerged as a representation of China in various contexts over time. Its use as a national symbol reflects less the constructed image a nation would want to portray generally, and instead the specific image that

China needed at particular historical moments both domestically and internationally. It has been an effective representation of China because, in spite of shifting contexts, the country has had a persistent need to appear more approachable, softer, or benign internally and externally. In other words, even though the communist government explicitly rejected being associated with the imperial-tainted image of the dragon, China continued to be seen as inscrutable, iron-handed, or threatening. China's need to portray panda traits varied and sometimes reflected opportunity as much as it did objective.

The panda is distinct from all other animals in the degree to which it is able to serve human emotional needs and thrive in the human imagination. It could be argued that it has survived extinction because it is cute. If so, humans then have a need for pandas apart from the contribution to their ecosystem. The panda flourishes because of humans, even though human settlement threatens panda habitat and existence. People have gone to great lengths to counter this human threat, however, because of the emotional value they place on the panda. As a result, there are panda reserves, rescue efforts, and robust fundraising efforts for its protection. The image of the panda that lives in the human imagination contributes directly to the animal's continued survival. That the panda is fluffy and cute matters as much with regard to its efficacy as a symbol as the perpetuation of the wild population. The dichotomy of the panda's existence as animal and image demonstrates humanity's need to connect to nature and its limited capacity to do so.

Because the giant panda is serendipitously exclusive to China, the PRC has been the fortunate host of this unusually appealing animal. The emergence of the giant panda from its shrouded existence in China's western mountains into the limelight of high diplomacy and universal recognition occurred with the convergence of the animal's innate characteristics, the political and economic transformations of China, and the affection of both the people of China and the global community. The giant panda as a symbol not only reflects the dynamic transformations that the PRC has undergone since its founding in 1949, but it also played a major role at each turn. One can trace the idealism of the early PRC in the efforts to use the panda to transform mass thought during the 1950s. The limits of nature that the Great Leap Forward highlighted and China's efforts to recover from that campaign were written across the landscape in panda country with new policies and new reserves during the early 1960s. The image of the giant panda demonstrated remarkable agility in dodging the political turmoil of the Cultural Revolution; it directly participated in China's strides toward self-reliance during the peak of this era as well as its march toward international engagement as the era closed during the 1970s. Ironically, as a species under the threat of starvation, the panda became a symbol of China's growing prosperity during the 1980s. The effort of China's population to save the panda was testament to measurable economic expansion. Since the 1990s, the burgeoning tourism industry and broadening scientific collaboration continue to beckon the world to China to visit the panda's hinterland home. The giant panda as animal and image has been instrumental in bridging nature and nationalism, as well as science and society. Difficult to define, impossible to ignore, and easy to underestimate, the giant panda is much like modern China. And like its home country, it is impossible to imagine a world without it.

Notes

INTRODUCTION

1. George B. Schaller, Hu Jinchu, Pan Wenshi, and Zhu Jing, *The Giant Pandas of Wolong* (Chicago: Chicago University Press, 1985); Pan Wenshi, Lü Zhi, Zhu Xiaojian, Wang Dajun, Wang Hao, Long Yu, Fu Dali, and Zhou Xin, *Jixu shengcun de jihui* (A chance for lasting survival). Beijing: Beijing Daxue chuban she, 2001; Davis, D. Dwight, *The Giant Panda: A Morphological Study of Evolutionary Mechanisms*. Chicago: Chicago Natural Museum, 1964, among many others.
2. Ruth Harkness, *The Lady and the Panda* (London: Nicholson and Watson, 1938); Ramona and Desmond Morris, *Men and Pandas* (New York: McGraw-Hill Book Company, 1966); Henry Nicholls, *The Way of the Panda: The Curious History of China's Political Animal* (New York: Pegasus Books, 2011), among others.
3. Mark Hineline, lecture, "The Built Environment," University of California San Diego, 2000.
4. Historical geographers include scholars such as Richard Louis Edmonds; economic and agricultural historians such as Peter Purdue early on integrated the environment in their analysis of Chinese history. Harriet Ritvo and others pioneered the field of animal studies. Historical studies of China such as Mark Elvin's *Retreat of the Elephants* on premodern China and Robert Marks' examination of late-imperial China in *Tigers, Rice, Silk, and Silt* are among other foundational works on the environmental history of China. These books focused on the symbolic power of the animals and used them as effective launching pads from which to delve into other historical questions.
5. Torah Kachur, "Is China the World's New Scientific Superpower?" *CBC News*, September 1, 2016, http://www.cbc.ca/news/technology/china-science-superpower-1.3743556, (Accessed January15, 2017).
6. W. D. Matthew and Walter Granger, "Fossil Mammals from the Pliocene of Szechuan, China," *Bulletin of the American Museum of Natural History* 48, no.17 (December 10, 1923): 579.

7. Schaller et al., *The Giant Pandas of Wolong*; Rui Peng, Bo Zeng, Xiuxiang Meng, Bisong Yue, Zhihe Zhang, and Fangdong Zou, "The Complete Mitochondrial Genome and Phylogenetic Analysis of the Giant Panda (*Ailuropoda melanoleuca*)," *Gene* 397 (2007): 76–83.
8. Fa-ti Fan, "Redrawing the Map: Science in Twentieth-Century China," *Isis* 98 (2007): 524–538.

CHAPTER 1

1. Hua Mei's father, Shi Shi, was exchanged for another male panda, Gao Gao, in January of 2003; James Steinburgh, "Hua Mei's Trip to China Postponed Again, This Time by SARS Fears; Human Companions Would Be Put at Risk," *Union Tribune* (San Diego), April 23, 2003.
2. Ian Markham-Smith, "US Zoo to Be Test Case for Panda Plan," *South China Morning Post* (Hong Kong), January 22, 1995.
3. The giant pandas in Mexico are an exception to this arrangement. The giant pandas that have lived in the Chapultepec Zoo in Mexico City were actual gifts from China and belonged to Mexico. The offspring, born in 1981, 1983, and 1990, were therefore offspring of pandas that Mexico owned. As such, China did not have any official jurisdiction over the baby pandas that the gift pandas produced. This did not prevent China from getting upset when the Mexican zoo offered one of its pandas to the Memphis Zoo on a short-term loan in exchange for several large primates (please see George B. Schaller, *The Last Panda* (Chicago: University of Chicago, 1993), 237 for more details on this specific event). Captive breeding of giant pandas was also successful in Madrid in 1982 and Tokyo in 1988.
4. Donald Harper, "The Cultural History of the Giant Panda (Ailuropoda melanoleuca) in Early China," *Early China* 35–36 (2012–2013): 186. The dragon and the phoenix are closely associated with the emperor and empress of China. The crane was considered to be generally auspicious. Craig Clunas, *Art in China* (Oxford, England: Oxford University Press, 1997), 56. Tigers were seen as connected with "the other world" and as a vehicle to it. Sara Allen, *The Shape of the Turtle: Myth, Art, and Cosmos in Early China* (New York: State University of New York Press, 1991), 154. The cicada was a symbol of rebirth and immortality. Please see Berthold Laufer's study, *Jade. A Study in Chinese Archeology and Religion* (New York: Dover Publishing, 1974).
5. Père David was interested in joining this order because they were sending missionaries to China, a country that intrigued him and his naturalist curiosities. Ramona Morris and Desmond Morris, *The Giant Panda* (London: Kogan Page, 1981), 25–27; Schaller, *Last Panda*; Keith Laidler and Liz Laidler, *Pandas* (London: BBC Books, 1992); Armand David, *Journal de mon troisieme d'exploration dans l'empire de Chinois par l'Abbé Armand David* [Diary of my explorations in the Chinese empire by Abbé Armand David], (Paris, Hachette, 1875); Helen M. Fox,

trans., *Abbé David's Diary: Being an Account of the French Naturalist's Journeys and Observations in China in the Years 1866–1869* (Cambridge, MA: Harvard University Press, 1949). Lazarist was the name given to the missionaries who served under the St. Vincent de Paul order because it was headquartered at the priory of Saint Lazare in Paris. Saint Lazare memorializes a poor man named Lazarus who suffered from leprosy and poverty as described in the parable from Luke 16: 19–25. St. Vincent de Paul was a French monk who had particular compassion for the poor.

6. Laidler and Laidler, *Pandas,* 38; Schaller, *Last Panda,* 132.
7. Laidler and Laidler, *Pandas,* 38.
8. David, *Abbé David's Diary,* 276–283; Armand David, "Journal d'u voyage dans le center de la Chine et dans le Thibet Oriental" [Diary of my travels to the center of China and in Tibet], *Bulletin Nouvelle Archives Muséum d'Histoire Naturelle de Paris* 10: 3–82; Also quoted in Morris and Morris, *Giant Panda,* 28, and Laidler and Laidler, *Pandas,* 45.
9. Schaller et al., *The Giant Pandas of Wolong,* 225; Morris and Morris, *Giant Panda,* 29.
10. Fox, *Abbé David's Diary,* 283.
11. Hu Jinchu along with Pan Wenshi and Zhu Jing collaborated with George B. Schaller in his seminal wild panda behavioral research conducted in Sichuan Province between 1980 and 1985. This project is described in great detail in the two books that the project produced: Schaller et al., *Giant Pandas of Wolong* and Schaller, *Last Panda.*
12. Hu Jinchu, *Da xiongmao de yanjiu* [Research on the giant panda], (Shanghai: Shanghai keji jiaoyu chuban she, 2001), 1; Zhou Jianren, "Guanyu xiongmao," *Renmin ribao* [People's Daily], July 6, 1956, 8; *Siku quanshu* [Complete books of the four treasuries], Wenyuange siku quanshu dianzi ban [electronic source], (Hong Kong: Dizhi wenhua chuban youxian gongsi, Zhongwen daxue chuban she, 1999).
13. Hu Jinchu, *Da xiongmao de yanjiu,* 3.
14. Harper, "Cultural History of the Giant Panda," 219.
15. Hu Jinchu, *Da xiongmao de yanjiu,* 1.
16. Harper, "Cultural History of the Giant Panda,"185.
17. Li Shizhen, *Bencao ganmu* [*Materia medica*], (Shanghai: Shanghai guji chuban she, 1991 [1596]).
18. Harper, "Cultural History of the Giant Panda," 186.
19. Harper, "Cultural History of the Giant Panda," 188.
20. Harper, "Cultural History of the Giant Panda," 192–195.
21. Harper, "Cultural History of the Giant Panda," 185–224.
22. Harper, "Cultural History of the Giant Panda," 187.
23. Harper, "Cultural History of the Giant Panda," 188.
24. Li Shizhen, *Bencao ganmu.*

25. *Gujin tushu jicheng* [Compilation of Books and Illustrations, Past and Present] (Taipei: Tingwen shu chu, 1977 [1725]).
26. E. H. Wilson, *A Naturalist in Western China* (London: Cadogan Books, 1913), 184.
27. Wilson, *Naturalist in Western China*, 183.
28. Theodore Roosevelt and Kermit Roosevelt, *Trailing the Giant Panda* (New York: Blue Ribbon Books, Inc., 1929), 2.
29. Roosevelt and Roosevelt, *Trailing the Giant Panda*, 3.
30. The brothers trekked in the region that they eventually succeeded in hunting the giant panda from March 6 through April 13, 1928. Roosevelt and Roosevelt, *Trailing the Giant Panda*, 145, 221, 225.
31. Two Chinese-American brothers and a number of local hunters assisted Ruth Harkness in this mission and captured a baby panda cub. Harkness found a home for the baby panda cub, Su Lin, at the Brookfield Zoo in Chicago. Su Lin lived in Chicago for about a year before he got sick and died. The Ruth Harkness expedition is also recounted in many other books, the most detailed is her own: Ruth Harkness, *The Lady and the Panda*, 187; Morris and Morris, *Men and Pandas*, 62–78.
32. Morris and Morris, *Giant Panda*, 88.
33. Schaller, *Last Panda*, 46.
34. Roosevelt and Roosevelt, *Trailing the Giant Panda*, 228. It has been suggested that the resistance to eating panda meat that the members of the Lolo ethnic minority group, now more commonly referred to as Yi—Lolo is now considered to be a derogatory term—expressed in this account may have been based on a religious belief or superstition. Please see Stevan Harrell, "The History of the History of the Yi," in *Cultural Encounters on China's Ethnic Frontiers*, ed. Stevan Harrell (Seattle: University of Washington Press, 1995). While the Roosevelts expressed surprise, they did not probe the issue further. Without any more evidence, it is difficult to know definitively why this group did not eat panda with the Roosevelt brothers. One can deduce, however, that the giant panda was not casually considered game for meat as far as this group of people was concerned.
35. Recorded by Schaller, who personally attended the trial, in Schaller, *Last Panda*, 126.
36. *Gujin tushu jicheng*, juan 670, xing 660. Also see Hu Jinchu, George B. Schaller, Pan Wenshi, and Zhu Jing, *Wolong de da xiongmao* [The giant pandas of Wolong] (Chengdu: Sichuan kexue jishu chuban she, 1985), 3–4.
37. Based on personal interviews with three members of the Baima community, Pingwu County, Sichuan, June 2002. It is difficult to historicize this belief. Government documents from the early 1960s indicate that Baima people did in fact hunt giant pandas, at least on occasion. Those documents probably refer to a local response to China's widespread famine following the Great Leap Forward.
38. Schaller, *Last Panda*, 227.
39. *Guangming ribao*, May 5, 1999; Sheryl WuDuun, "Pessimism Is Growing on Saving Pandas From Extinction," *New York Times*, June 11, 1991, C4. According

to personal interviews conducted in China 2001–2002, six people received the death penalty for poaching giant pandas. While this law has not been officially removed from the books, no one has reportedly been sentenced to death for this offense in recent years.

40. "Woman Accused of Peddling Panda Pelt," *USA Today*, July 23, 2008.
41. Wilson, *Naturalist in Western China*, 183. *Peh Hsiung* is Ernest Wilson's romanization of *bai xiong*, the local name for the giant panda.
42. Schaller et al., *Giant Pandas of Wolong*, 129.
43. Schaller et al., *Giant Pandas of Wolong*, 225, Morris and Morris, *Giant Panda*, 30.
44. D. Davis, *Giant Panda: A Morphological Study*, 14–15.
45. Davis, *Giant Panda: A Morphological Study*, 14.
46. More recently the term "red panda" has been promoted and favored over "lesser panda" in the spirit of finding alternatives to the seemingly demeaning categorization of "lesser" as well as using a term closer to the Latin name.
47. Hu Jinchu, *Da xiongmao de yanjiu*, 2.
48. Pan Wenshi, Gao Zhengsheng, Lü Zhi, Xia Zhengkai, Zhang Miaodi, Ma Cailing, Meng Guangli, Zhe Xiaoye, Lin Xuzhuo, Cai Haiting, and Chen Fengxiang, *Qinling da xiongmao de ziran bihu suo* [The giant panda's natural refuge in the Qinling Mountains] (Beijing: Beijing Daxue chuban she, 1988), 17.
49. Davis, *Giant Panda: A Morphological Study*, 14–15.
50. Rui Peng, Bo Zeng, Xiuxiang Meng, Bisong Yue, Zhihe Zhang, and Fangdong Zou, "The Complete Mitochondrial Genome and Phylogenetic Analysis of the Giant Panda (*Ailuropoda melanoleuca*)," *Gene* 397 (2007): 76–83. Schaller et al., *Giant Pandas of Wolong*, 228; Schaller, *Last Panda*, 266–267.
51. Davis, *Giant Panda: A Morphological Study*, 16. For further detail, see E. R. Lankester, "On the Affinities of *Æluropus malanoleucus*," *Linnean Society of London* 2, no. 8 (1901): 163–171.
52. Matthew and Granger, "Fossil Mammals from the Pliocene," 579. To aid with the Latin terminology, the *Aeluropus* refers to *Aeluropoda* or *Ailropoda*, the giant panda genus. Hyaenarctos is an extinct separate branch of the bear evolutionary stem. *Ursidae* is the bear family. The use of terms such as "stem" and "family" adhere to the Linnaean taxonomical system rather than the more recent system of cladistics because the authors of this quotation were working within the context of the Linnaean system. According to the 2012 adjustments to the Geological Society of America geologic time table, the periodization of the Pliocene epoch was 2.6 to 5.3 million years ago. The concept that something could be "modernized" in the Pliocene implies that relatively minor evolutionary change has occurred in these animals since the Pliocene. https://www.geosociety.org/GSA/Education_Careers/Geologic_Time_Scale/GSA/timescale/home.aspx.
53. Davis, *Giant Panda: A Morphological Study*, 16.
54. Stephen J. Gould, "The Panda's Thumb," *The Panda's Thumb: More Reflections in Natural History* (New York: W.W. Norton and Company, 1980), 22.

55. Davis, *Giant Panda: A Morphological Study*, 326.
56. The general trend toward typing the giant panda as a bear can be traced in scientific literature and, perhaps even more revealing, in reference lists that tap the scientific literature for information such as the IUCN Red List of endangered species. Please see: IUCN (International Union for the Conservation of Nature), Red List of Threatened Species http://www.iucnredlist.org/search/details.php/712/all; CITES (Convention on Trade in Endangered Species), Appendix I http://www.cites.org/eng/app/appendices.shtml; Tetsuo Hashimoto et al., "The Giant Panda Is Closer to a Bear, Judged by α- and β-Hemoglobin Sequences," *Journal of Modern Evolution* 36 (1993): 282–289.
57. Schaller, *Last Panda*, 262.
58. Rui et al., "Phylogenetic Analysis of the Giant Panda," 82.
59. Olaf R. P. Bininda-Emonds, "Phylogenetic Position of the Giant Panda: Historical Consensus through Supertree Analysis," in *Giant Pandas: Biology and Conservation*, ed. Donald Lindburg and Karen Baragona (Berkeley: University of California Press, 2004), 14, 289.
60. Davis, *Giant Panda: A Morphological Study*, 15.
61. Pan et al., *Qinling da xiongmao*, 3–21; Chris Catten, *Pandas* (London: Christopher Helm, 1990), 31.
62. Davis, *Giant Panda: A Morphological Study*, 17; Zhu Jing and Long Zhi, "Da xiongmao de xingshuai [The rise and fall of the giant panda]," *Acta Zoologica Sinica* 29, no.1 (March 1983): 93–103.
63. "'Giant Panda Originates in Europe' Incredible [*sic*]," *Beijing Review*, 41, no. 22 (June 1–7, 1998): 28.
64. Robert M. Hunt, "A Paleontologist's Perspective on the Origin and Relationships of the Giant Panda," in *Giant Pandas: Biology and Conservation*, ed. Donald Lindburg and Karen Baragona (Berkeley: University of California Press, 2004), 48.
65. Hunt, "A Paleontologist's Perspective," 48; The Miocene epoch occurred 5.3–23 million years ago according to the Geological Society of America.
66. Changzhu Jin, Russell L. Ciochon, Wei Dong, Robert M. Hunt, Jr., Marc Jaeger, and Qizhi Zhu, "The First Skull of the Earliest Giant Panda," *Proceedings of the National Academy of Sciences* 104, no. 26 (June 26, 2007): 10932–10937; Sindya N. Bhanoo, "Ancient Bear May Be Ancestor of Giant Panda," *New York Times*, November 19, 2012.
67. Matthew and Granger, "Fossil Mammals from the Pliocene," 579.
68. Stephen Jay Gould, "How Does the Panda Fit?" in *An Urchin in the Storm* (New York: W.W. Norton and Company, 1987), 23.
69. Gould, "The Panda's Thumb," 23.
70. Changzhu Jin et al., "The First Skull," 10936.
71. Russell L. Ciochon, Professor and Chair, Department of Anthropology, University of Iowa, and author of aforementioned study of panda fossils, personal correspondence, December 10 and November 21, 2008.

72. Changzhu Jin et al., "The First Skull," 10932.
73. Shuping Yao, "Chinese Intellectuals and Science: A History of the Chinese Academy of Sciences (CAS)," *Science in Context* 3, no. 2 (1989): 448.
74. The Office of Naval Research (ONR) and the Atomic Energy Commission (AEC) are among the many organizations in the United States that connect these three forces. For more examples and further detail, see Everett Mendelsohn, Merritt Roe Smith, and Peter Weingart, *Science Technology and the Military* (Dordrecht: Kluwer Academic Publishers, 1988). A significant distinction between the United States' and China's efforts to enlist science under the service of the state is that in the case of China, this correlation was an integral part of the nation-building plan at its inception.
75. Shuping Yao, "Chinese Intellectuals and Science," 449.
76. Zhou Jianren was the younger brother of the famous literary figure Lu Xun. Zhou also worked as a translator and worked with Ye Duzhuang on the translation of Charles Darwin's *The Origin of Species.*
77. Zhou Jianren, "Guanyu xiongmao," 8.
78. Zhou Jianren, "Guanyu xiongmao," 8.
79. Zhou Jianren, "Guanyu xiongmao," 8.
80. Zhou Jianren, "Guanyu xiongmao," 8. It should be noted here that although the panda diet is 99% bamboo, pandas actually do (very) occasionally eat meat. However, at the time of this article, it was thought that they were strict herbivores trapped in a carnivore's body.
81. Zhou Jianren, "Guanyu xiongmao," 8.
82. Roderick MacFarquhar, *The Origins of the Cultural Revolution, vol. 1, Contradictions Among the People, 1956–1957* (New York: Columbia University Press, 1974), 48–52.
83. I would like to thank Ye Wa, Ph.D. for pointing me toward the notion of the scientific movement.
84. *Renmin ribao* [*People's Daily*] (hereafter, RMRB), November 28, 1963, 2.
85. RMRB, November 28, 1963, 2. "Liberation" refers to the Chinese Communist Party's victory in taking power over China in 1949.
86. RMRB, November 28, 1963, 2.
87. Zhu Jing and Li Yangwen, *Da xiongmao* [Giant panda] (Beijing: Science Press, 1980), Appendix.
88. Zhu and Li, *Da xiongmao,* Appendix.
89. *Liang you* [Young companion] 153, no. 31 (April 1940), 42.
90. Guo An, "Wo guo zui da de dongwu yuan–Beijing dongwu yuan," *Renmin ribao,* May 6, 1956, 3; Zhou Jianren, "Guanyu xiongmao," 8.
91. Nicholas Menzies, *Forest and Land Management in Imperial China* (New York: St. Martin's Press, 1994); Susan Fernsebner, "Material Modernities: China's Participation in World's Fairs and Expositions, 1876–1955" (PhD diss., University of California, San Diego, 2002).
92. Zhou Jianren, "Guanyu xiongmao," 8.

93. Guo An, "Wo guo zui da de dongwu yuan," 3.
94. RMRB, November 28, 1963, 2.
95. Stephan Grauwels, "Taiwan Rejects China's Offer of Pandas," *Associated Press Online*, March 31, 2006 http://www.lexisnexis.com.ezproxy1.lib.ou.edu/us/lnacademic/results/docview/docview.do?docLinkInd=true&risb=21_T5500773876&format=GNBFI&sort=BOOLEAN&startDocNo=1&resultsUrlKey=29_T5500773879&cisb=22_T5500773878&treeMax=true&treeWidth=0&csi=147876&docNo=1 (Accessed January 9, 2009).
96. Grauwels, "Taiwan Rejects Pandas," 31 March 2006.

CHAPTER 2

1. IUCN Red List of Threatened Species.
2. Sichuan Forest Department Representative, Chengdu, Sichuan Province, PRC, interview by author, January 2002; Geologist, Peking University, Beijing, interview by author, March 2002.
3. Shuping Yao, "Chinese Intellectuals and Science," 130.
4. Zheng Zuoxin, "Sulian dongwu qu xi yu shoulie yanjiu shiye de fazhan jinkuang" [Recent status of the development of Soviet animal reserves and hunting research industries], *Dongwuxue zazhi* [Journal of zoology] 3, no. 1 (1959), 31; Suzanne Pepper, "Learning for the New Order," *The Cambridge History of China*, Volume 14, *The People's Republic, Part I: The Emergence of Revolutionary China, 1949–1965*, ed. Roderick MacFaraquhar and John K. Fairbank (Cambridge: Cambridge University Press, 1987), 197–203; Shuping Yao, "Chinese Intellectuals and Science," 451–453.
5. Douglas Weiner, among others, has charted the grave pitfalls that befell these fields, sometimes with catastrophic results, due to the political subjugation of science in the USSR. Weiner also traced the struggles, resistance, and successes of scientists who fought the misguided directions in which politics sometimes pulled science. Please see Douglas R. Weiner, *Models of Nature: Ecology, Conservation, and Cultural Revolution in Soviet Russia* (Pittsburgh: University of Pittsburgh Press, 1988) and Douglas R. Weiner, *A Little Corner of Freedom: Russian Nature Protection from Stalin to Gorbachëv* (Berkeley: University of California Press, 1999).
6. Lysenko was a Soviet theorist of heredity. For a succinct and seminal study on him and policies associated with him, see David Joravsky, *The Lysenko Affair* (Chicago: Chicago University Press, 1986 [1970]). For an examination of his impact on China, see Laurence Schneider, "Editor's Introduction: Lysenkoism in China: Proceedings of the 1956 Qingdao Genetics Symposium," *Chinese Law and Government* 19, no. 2 (1986); Schneider, *Biology and Revolution in Twentieth-Century China*, 115–212.
7. N. Laluosikaya, "Ziran jieli meiyou mimi" [Nature has no secrets], *Zhishi jiushi liliang* [*Knowledge is Power*] (1958), 17. This is the romanization of the Chinese transliteration of the name of this Soviet author.

8. Wu Zhonglun, ed., *Zhongguo ziran baohu qu* [China's nature reserves] (Shanghai: Keji jiaoyu chuban she, 1996), 35. Dinghu Shan reserve is located in Zhaoqing County, Guangdong Province.
9. Wu Zhonglun, *Zhongguo ziran baohu qu,* 539.
10. Zhu Kezhen, "Kaizhan ziran baohu gongzuo" [Developing nature protection work], *Zhu Kezhen wen ji* [Collected essays of Zhu Kezhen], 436. The information that the purpose of setting up a reserve in a colder climate was for comparative study comes from a Sichuan Forest Department Representative, interview, January 2002.
11. Weiner, *Little Corner of Freedom,* 210.
12. Zhu Kezhen, "Zhongguo ziran qu hua (chugao) 'xu'" [Preface of China's nature reserve plan, draft], *Zhu Kezhen wen ji* [Collected essays of Zhu Kezhen] (Beijing: Kexue chuban she, 1959), 377.
13. Zhu Kezhen, "Zhongguo ziran," 377.
14. Jin Jieliu, "Guanyu 'ziran baohu'" [Concerning 'nature protection'], 动物学杂志*Dongwuxue zazhi* [Journal of zoology] 1, no. 3 (1957), 129.
15. Jin Jieliu, "Guanyu 'ziran baohu,'" 129–131.
16. G. P. Dementiev, "Sulian de ziran baohu" [Nature protection in the Soviet Union], *Dongwuxue zazhi* [Journal of Zoology] 1, no. 4 (1957): 210–211. Please note that in the text of this chapter I follow Douglas Weiner's spelling of this author's name (Dement'ev) and of the names of other Soviet figures. The Chinese source romanizes the name as it appears on this footnote.
17. Dementiev, "Sulian de ziran baohu," 210–211.
18. Jin Jieliu, "Guanyu 'ziran baohu,'" 129.
19. Zhu Kezhen, "Zhongguo ziran," 377.
20. Zuoye Wang, "Zhu Kezhen," in *Complete Dictionary of Scientific Biography,* ed. Noretta Koertge (New York: Charles Scribner's Sons, 2007).
21. Zhu Kezhen, "Zhongguo ziran," 377.
22. Zhu Kezhen, "Zhongguo ziran," 377.
23. Samuel P. Hays, *Conservation and the Gospel of Efficiency: The Progressive Conservation Movement, 1890–1920* (New York: Atheneum, 1979 [1959]), 2.
24. Zuoye Wang, "Zhu Kezhen."
25. Judith Shapiro, *Mao's War Against Nature: Politics and the Environment in Revolutionary China* (Cambridge, England: Cambridge University Press, 2001), 67–94.
26. Penny Kane, *Famine in China 1959–1961: Demographic and Social Implications* (Hong Kong: The Macmillan Press, 1988), 88–91; Jasper Becker, *Hungry Ghosts: Mao's Secret Famine* (New York: Henry Holt and Company, 1996), 270.
27. PRC, Guowu yuan [State Council], "Guowu yuan guanyu jiji baohu he heli liyong yesheng dongwu ziyuan de zhishi" [State Council notice concerning the active protection and rational use of wild-animal resources], (Guolin Tanzi 287 hao), September 14, 1962, 1.
28. PRC, Guowu yuan (Guolin Tanzi di 287 hao), September 14, 1962, 3.

29. PRC, Guowu yuan (Guolin Tanzi di 287 hao), September 14, 1962, 1–4.
30. PRC, Guowu yuan [State Council], "Guowu yuan dui linye bu guanyu kaizhan wo guo shoulie shiye baogao de pifu" [Report on the State Council address to the Ministry of Forestry concerning the development of our national hunting industry] (1958), *Guanyu yesheng dongwu ziyuan de wenjian ji* [Collection of documents concerning wild-animal resources], ed. Lin zheng baohu si [Forest Protection Department] (Beijing: Linye bu, 1987), 13.
31. PRC, Guowu yuan [State Council], "Zhonghua renmin gonggong he linye bu wei tingzhi shengchan he bianyong shi yong pu shou wan de lianhe tongzhi" [Announcement to the People's Government and the Ministry of Forestry calling for the end of the production and sale of specific animal poison] (1961), in *Guanyu yesheng dongwu ziyuan de wenjian ji* [Collection of documents concerning wild-animal resources], ed. Lin zheng baohu si [Forest protection department] (Beijing: Linye bu, 1987), 100.
32. Dementiev, "Sulian de ziran baohu," 210–211.
33. Zheng Zuoxin, "Sulian dongwu qu," 31.
34. Dementiev, "Sulian de ziran baohu," 210.
35. Shaanxi sheng, Linye ting [Shaanxi Province, Forestry department], "Shaanxi sheng shoulie shiye guanli zanxin banfa" [Shaanxi Province announcement concerning the issuing of temporary methods of hunting industry management for Shaanxi Province], (Huilin Li zi di 793 hao), December 26, 1962.
36. PRC, Guowu yuan (Guolin Tan zi di 287 hao), September 14, 1962, 5.
37. Dementiev, "Sulian de ziran baohu," 210–211.
38. PRC, Guowu yuan (Guolin Tan zi di 287 hao), September 14, 1962, 3.
39. Zheng Zuoxin, "Sulian dongwu qu," 32.
40. Dementiev, "Sulian de ziran baohu," 211; Zheng Zuoxin, "Sulian dongwu qu," 31.
41. Weiner, *Little Corner of Freedom*, 389.
42. PRC, Guowu yuan (Guolin Tan zi di 287 hao), September 14, 1962, 2–5.
43. Nicholas R. Lardy, "Economic Recovery, 1963–1965," *The Cambridge History of China, Volume 14, The People's Republic, Part I: The Emergence of Revolutionary China, 1949–1965*, ed. Denis Twitchett and John K. Fairbank (Cambridge: Cambridge University Press, 1987), 378–397.
44. PRC, Guowu yuan, Linye bu [State Council and Ministry of Forestry], "Guanyu guoying linchang jingying shoulie shiye de ji xiang guiding" [Regulations concerning the operation and management of the hunting industry in national forest areas] (1961), *Guanyu yesheng dongwu ziyuan de wenjian ji* [Collection of documents concerning wild-animal resources], ed. Lin zheng baohu si [Forest protection department] (Beijing: Linye bu, 1987), 107–112; Shaanxi sheng, Linye ting (Huilin Li zi di 793 hao), December 26, 1962.
45. PRC, Linye bu (Ministry of Forestry), "Guanyu guoying linchang jingying guanli shoulie shiyde ji xiang guiding" (Concerning the Operation and Management of

Hunting Industry Regulations in National Forest centers), Lin jing Luo zi di 145 hao (10 May 1962), 107.

46. PRC, Guowu yuan (Guolin Tan zi di 287 hao), September 14, 1962, 3.
47. PRC, Guowu yuan (Guolin Tan zi di 287 hao), September 14, 1962, 1, 4.
48. PRC, Guowu yuan (Guolin Tan zi di 287 hao), September 14, 1962, 3.
49. PRC, Guowu yuan (Guolin Tan zi di 287 hao), September 14, 1962, 3.
50. James C. Scott, *The Moral Economy of the Peasant: Rebellion and Subsistence in Southeast Asia* (New Haven: Yale University Press, 1977).
51. PRC, Guowu yuan, "Guoying linchang jingying shoulie shiye," (1961), 110.
52. The Chinese government did not advertise its grain importation. This measure was desperate and its publicity would force the acknowledgment that not only was China unable to recover on its own, but the exorbitant grain taxes and export of grain during the previous years had actually caused the famine.
53. Dong Zhiyong, ed., *Zhongguo linye nianjian* [China Forestry Yearbook], (Beijing: Zhongguo linye chuban she, 1987), 80.
54. Jin Jieliu, "Guanyu 'ziran baohu,'" 130.
55. PRC, Guowu yuan, (Guolin Tan zi di 287 hao), September 14, 1962, 2.
56. Jin Jianming, ed., *Ziran baohu gailun* [An introduction to nature protection], (Beijing: Zhongguo huanjing kexue chuban she, 1991), 168.
57. PRC, Guowu yuan (Guolin Tan zi di 287 hao), September 14, 1962, 4.
58. *Gujin tushu jicheng*, juan 660, xing 670; Schaller et al., *The Giant Pandas of Wolong*, 3–4.
59. Roosevelt and Roosevelt, *Trailing the Giant Panda*, 228; Schaller, *The Last Panda*, 126.
60. Kang Jia (pseudonym), Baima woman, Pingwu County, Sichuan, PRC, interview by author, May 29, 2002.
61. Sichuan sheng, Pingwu xian, Linye ju (Pingwu county forestry bureau), "Pingwu xian shoulie gongzuo kaizhan qingkuang jianjie" (A concise report on the status of the development of work on hunting in Pingwu county) (Sept. 13, 1965), 2–4.
62. PRC, Guowu yuan, "Guoying linchang jingying shoulie shiye," (1961), 108; John F. Reiger, *American Sportsmen and the Origins of Conservation* (New York: Winchester Press, 1975), 36.
63. Weiner, *Models of Nature*, 30; Weiner, *Little Corner of Freedom*, 204–205, 386–387. Predation control predates Soviet conservation and was a debated issue from Lenin's tenure through the 1980s; Thomas Dunlap, *Saving America's Wildlife* (Princeton: Princeton University Press, 1988), 39–40, 113. During the 1920s, the United States opened a special office, Predator and Rodent Control (PARC), precisely for the sake of dealing with this issue. Activists and scientists debated the extermination of predators in the United States from the founding of PARC forward. The United States was even more aggressive than China in its early predator control campaigns. Not only were such questionable methods as poison

utilized, but the US government also sent out teams of hunters to cleanse the hills of coyotes and wolves.

64. Tang Tansheng, "Chuan bei jiu xian de maopi shou ji qi liyong de chubu baogao" [Report on the early steps in finding pelt animals and other useful items in the northern nine counties of Sichuan], *Dongwuxue zazhi* [Journal of Zoology] 4, no. 5 (1960), 197.
65. Tang Tansheng, "Chuan bei jiu xian," 197.
66. PRC, Guowu yuan, (Guolin Tan zi di 287 hao), September 14, 1962, 4; PRC, Guowu yuan, "Guoying linchang jingying shoulie shiye," (1961), 108.
67. Zhu Jing. "Jiefang yihou wo guo de shoulie, xunyang, yu baohu gong zuo" (National hunting, rearing, and nature protection work since liberation). *Dongwuxue zazhi* (Journal of zoology) 6, no. 6 (1964), 315–316.
68. Zhu Jing, "Jiefang yihou wo guo de shoulie," 315–316.
69. Zhu Jing, "Jiefang yihou wo guo de shoulie," 316.

CHAPTER 3

1. In 1958 the State Council assigned the Ministry of Forestry and its subordinate offices jurisdiction over all hunting regulation. This 1962 State Council directive largely focused on the regulation of hunting. Hu Tieqing, former head of Sichuan Provincial Wild Animal Protection Department, interview by author, Chengdu, Sichuan Province, PRC, January 14, 2002.
2. Feng Yunwu, former deputy director of Wanglang Nature Reserve, interview by author, Pingwu County, Sichuan Province, PRC, October 18, 2001.
3. PRC, Guowu yuan (Guolin Tanzi di 287 hao), September 14, 1962, 3–5; Sichuan sheng, Renmin weiyuan hui pizhuan sheng linye ting (Sichuan Provincial people's committee transmission to the provincial department of forestry). "Guanyu jiji baohu he heli liyong yesheng dongwu ziyuan de baogao" (Report concerning the active protection and rational use of wild animal resources) (Chuan nong zi 0191 hao) April 2,1963, 1.
4. Sichuan sheng, Renmin weiyuan hui (Chuan nong zi 0191 hao), April 2, 1963, 1.
5. Fu Yingquan, ed., *Sichuan zhuan* [Sichuan volume], Zhongguo ziran ziyuan cong shu [Chinese natural resources series], no. 33 (Beijing: Zhongguo huanjing kexue chuban she, 1995), 389–395.
6. Zhong Zhaomin, founder of Wanglang. Interview by author, Mianyang, Sichuan Province, PRC, January 22, 2002.
7. Sichuan sheng, Pingwu xian, Renmin weiyuanhui [Sichuan Province, Pingwu County, People's committee], "Guanyu jianli Wanglang ziran baohu qu de baogao" [Concerning the establishment of Wanglang Nature Reserve] (Hui ban zi di 022 hao), July 1, 1965, 1, 2.
8. Zhong Zhaomin. "Wanglang zai qianjin" (Wanglang forging forward), unpublished papers, April 2004.

9. Sichuan sheng, Renmin weiyuan hui (Chuan nong zi 0191 hao), April 2, 1963, 1.
10. Sichuan sheng, Renmin weiyuan hui (Chuan nong zi 0191 hao), April 2, 1963, 1.
11. Sichuan sheng, Renmin weiyuan hui (Chuan nong zi 0191 hao), April 2, 1963, 1.
12. Sichuan sheng, Renmin weiyuan hui (Chuan nong zi 0191 hao), April 2, 1963, 1.
13. There is some discrepancy in the founding dates of each of these reserves. The other three giant panda reserves are credited with being founded the same year as Wanglang by a few lists, yet most authorities interviewed named Wanglang as the first to be established. Hu Jinchu maintains that the Wolong reserve in Wenchuan County was established in 1963 after he discovered that the area contained giant pandas in his 1963 surveys. Wolong staff assert that the reserve was first established as a forest area and only in 1975 was it named a panda reserve with an expanded area. Wolong staff also note that it was not originally known that giant pandas lived in the area. This information conflicts with the copy of the Sichuan sheng, Renmin weiyuan hui (Chuan nong zi 0191 hao), April 2, 1963 document in my possession that specifically recommends the creation of a panda reserve in Wolong in 1963. The other official documents that I have regarding the creation of Wolong, however, assert the establishment of a forest area in 1965 and a panda reserve in 1975.
14. Sichuan sheng, Renmin weiyuan hui (Chuan nong zi 0191 hao), April 2, 1963, 2.
15. Sichuan sheng, Pingwu xian, Renmin weiyuan hui (Hui ban zi di 022 hao), July 1, 1965, 1, 2.
16. He Daihua, ed., *Ke'ai de jia xiang Pingwu* [Our endearing hometown of Pingwu] (Mianyang: Mianyang shi weicheng caiying chang ying shua, 1998), 2.
17. Hu Tieqing, interview, January 14, 2002; Feng Yunwu, interview, October 18, 2001; Zhong Zhaomin, interview, January 22, 2002.
18. Xie Zhong and Jonathan Gipps, *The 2001 International Studbook for Giant Panda* [*sic*] (*Ailuropoda melanoleuca*) (Beijing and London: Chinese Association of Zoological Gardens and Bristol Zoo Gardens, 2001), 4–22.
19. Qian Danning, ed., *Pingwu xian zhi* [Pingwu County gazetteer] (Chengdu: Sichuan kexue jishu chuban she, 1997), 118.
20. Sichuan sheng, Pingwu xian, Renmin weiyuan hui (Hui ban zi di 022 hao), July 1, 1965, 1.
21. Sichuan sheng, Pingwu xian, Renmin weiyuanhui, (Hui ban zi di 022 hao), July 1, 1965, 1.
22. Feng Yunwu, interview, October 18, 2001.
23. Zhong Zhaomin, "Wanglang zai qianjin," 2.
24. Zhong Zhaomin, "Wanglang zai qianjin," 2.
25. Feng Yunwu, interview, October 18, 2001.
26. Zhong Zhaomin, interview, January 22, 2002; Feng Yunwu, interview, October 18, 2001; Cai Li [pseudonym], former surveyor and staff member in Wanglang reserve, interview by author, Pingwu County, Sichuan Province, PRC, October 19, 2001; Geng Shengwa, former surveyor and member of the Baima community,

interview by author, Pingwu County, Sichuan Province, PRC, May 29, 2002; Geng Shengwa, Cai Li, and Chai Ningzhu are pseudonyms.

27. Zhong Zhaomin, interview, January 22, 2002.
28. Sichuan sheng, Pingwu xian, Renmin weiyuanhui (Hui ban zi di 022 hao), July 1, 1965, 1.
29. Sichuan sheng, Pingwu xian, Renmin weiyuan hui (Hui ban zi di 022 hao), July 1, 1965, 1.
30. Sichuan sheng, Linye ting (Sichuan provincial department of forestry), "Guanyu ziran baohu qu gongzuo de chubu yijian" (Suggestions on the introduction of nature reserve work) (Lin jing zi di 010 hao), February 17, 1965, 1.
31. Sichuan sheng, Linye ting (Lin jing zi di 010 hao), February 17, 1965, 1.
32. Sichuan sheng, Pingwu xian, Renmin weiyuan hui (Hui ban zi di 022 hao), July 1, 1965, 1. One hectare is equal to 2.45 acres.
33. Sichuan sheng, Pingwu xian, Renmin weiyuan hui (Hui ban zi di 022 hao), July 1, 1965, 2.
34. For example: Chen Guidi and Wu Chuntao, *Will the Boat Sink the Water: The Life of China's Peasants* (New York, Public Affairs, 2007); Yang Su, "Mass Killings in the Cultural Revolution: A Study of Three Provinces," in *The Chinese Cultural Revolution as History,* ed. Joseph W. Esherick, Paul G. Pickowicz, and Andrew Walder (Stanford: Stanford University Press, 2006).
35. Zhong Zhaomin, interview, May 27, 2002.
36. Sichuan sheng, Pingwu xian, Renmin weiyuan hui (Hui ban zi di 022 hao), July 1, 1965, 1.
37. Schaller et al., *The Giant Pandas of Wolong,* 254.
38. Zhong Zhaomin, interview, May 27, 2002.
39. Zhong Zhaomin, interview, May 27, 2002.
40. Qian Danning, *Pingwu xian zhi,* 515.
41. Qian Danning, *Pingwu xian zhi,* 515.
42. Qian Danning, *Pingwu xian zhi,* 515; Xiao Youyuan, *Pingwu Baima Zang zu* [The Baima Tibetans in Pingwu County] (Mianyang: Mianyang hong guang qiye yingshua chang, 2001), 49.
43. Local observer, Pingwu county, Sichuan, PRC, interview by author, October 2001.
44. Qian Danning, *Pingwu xian zhi,* 515–516.
45. Qian Danning, *Pingwu xian zhi,* 516.
46. Qian Danning, *Pingwu xian zhi,* 516.
47. Qian Danning, *Pingwu xian zhi,* 516; by 1964 the lumber company was called the Mianyang Prefecture Lumber Mill [*Mianyang zhuanqu famu chang* 绵阳专区伐木厂].
48. Qian Danning, *Pingwu xian zhi,* 517.
49. Baima villager, Pingwu County, Sichuan, PRC, interview by author, May 29, 2002.
50. Baima villager, interview, May 29, 2002.
51. Baima villager, interview, May 29, 2002.

52. Baima villager, interview, May 29, 2002.
53. Qian Danning, *Pingwu xian zhi*, 517.
54. Sichuan sheng, Pingwu xian, Renmin weiyuanhui (Hui ban zi di 022 hao), July 1, 1965, 1.
55. Integrated Conservation Development Program (ICDP), "Pingwu xian da xiongmao qixidi zonghe baohu yu fazhan xiangmu, Baima zang zu xiang shehui jingji diaocha baogao" (Pingwu county giant panda habitat Integration of Conservation and Development Program, Baima Tibetan community social and economic survey report). June 1998, 59–60.
56. Xiao Youyuan, 48.
57. Baima villager 2, Pingwu County, Sichuan, PRC, interview by author, May 29, 2002.
58. Zeng Weiyi, "Baima ren zuyu yanjiu jian jie," [A simple introduction of research on the classification of the Baima people] in *Baima zangzu yanjiu wenji* [Collected work on research on the Baima Tibetans], ed. Zeng Weiyi (Chengdu: Sichuan sheng minzu yanjiu suo, 2002), 209; Katia Chirkova, "Between Tibetan and Chinese: Identity and Language in Chinese South-West," *Journal of South Asian Studies* 30, no. 3 (2007): 415.
59. Sichuan sheng, Renmin weiyuan hui (Chuan nong zi 0191 hao), April 2, 1963, 1.
60. Sichuan sheng, Pingwu xian, Linye ju, "Pingwu xian shoulie gongzuo kaizhan qingkuang jianjie" (1965), 2.
61. Sichuan sheng, Pingwu xian, Linye ju, "Pingwu xian shoulie gongzuo" (1965), 2.
62. Sichuan sheng, Pingwu xian, Linye ju, "Pingwu xian shoulie gongzuo" (1965), 2.
63. PRC, Guowu yuan, (Guolin Tan zi di 287 hao), September 14, 1962, 3. The 1962 State Council policy demanded the active protection of wild animal resources and "strictly prohibited" the hunting of the giant panda among a select list of other species indigenous only to China.
64. Sichuan sheng, Renmin weiyuan hui (Chuan nong zi 0191 hao), April 2, 1963, 1.
65. Sichuan sheng, Renmin weiyuan hui (Chuan nong zi 0191 hao), April 2, 1963, 3.
66. Kang Jia [pseudonym], Baima woman, interview by author, Pingwu County, Sichuan, PRC, May 29, 2002.
67. Sichuan sheng, Renmin weiyuan hui (Chuan nong zi 0191 hao), April 2, 1963, 3.
68. Sichuan sheng, Pingwu xian, Linye ju, "Pingwu xian shoulie gongzuo" (1965), 2.
69. The report uses the term 战斗大队, which I translate as "combat forces." Pingwu xian, Linye ju, "Pingwu xian shoulie gongzuo" (1965), 6.
70. The referencing of ethnic minorities in China as barbarians and the former use of animal and insect radicals in the terms used to name various groups is broadly known. Stevan Harrell notes that ethnic minorities were viewed on a scale of cultural civility by the Han, some being more civilizable than others. To read further, please see Stevan Harrell, "Introduction," in *Cultural Encounters on China's Ethnic Frontiers*, ed. Stevan Harrell (Seattle: University of Washington Press, 1996), 9.

71. Sichuan sheng, Pingwu xian, Linye ju, "Pingwu xian shoulie gongzuo" (1965), 6; Ye Wa, Beijing resident during early 1960s, San Diego, CA, USA, interview by author, September 25, 2003.
72. Sichuan, Pingwu xian, Linye ju, "Pingwu xian shoulie gongzuo" (1965), 7.
73. Sichuan sheng, Pingwu xian, Linye ju, "Pingwu xian shoulie gongzuo" (1965), 7.
74. Sichuan sheng, Pingwu xian, Linye ju, "Pingwu xian shoulie gongzuo" (1965), 6.
75. Zhong Zhaomin, interview, May 27, 2002.
76. Baima villagers, interview by author, May 30, 2002.
77. See Kenneth Lieberthal and Michel Oksenberg, *Policy Making in China: Leaders, Structures, and Processes* (Princeton: Princeton University Press, 1988), 350; they cite David Goodman, *Centre and Province in the PRC: Sichuan and Guizhou, 1955–1965* (Cambridge, Cambridge University Press, 1986); Lardy, *Economic Growth and Distribution in China* (Cambridge, Cambridge University Press, 1978); Victor Faulkenheim, "Continuing Central Predominance," *Problems of Communism* 21, no. 4 (1972): 75–83.
78. Zhong Zhaomin, interview, January 22, 2002.
79. Cai Li, interview, October 19, 2001.
80. Lieberthal and Oksenberg, *Policy Making in China*, 351.

CHAPTER 4

1. Pingwu County is nearly 1,000 miles from China's capital, Beijing.
2. Wanglang ziran baohu qu da xiongmao diaocha zu [Wanglang Nature Reserve giant panda survey team], "Sichuan sheng Pingwu xian Wanglang ziran baohu qu da xiongmao de chubu diaocha" [Preliminary survey of giant pandas in the Wanglang Nature Reserve in Pingwu County, Sichuan] *Dongwuxue bao* [Acta Zoologica Sinica] 20, no. 2 (June 1974): 162. The Ministry of Agriculture and Forestry was a reorganization of the formerly separate ministries.
3. Wu Zhonglun, ed., *Zhongguo ziran baohu qu* [China's nature reserves] (Shanghai: Keji jiaoyu chuban she, 1996); Jin Jianming, Wang Liqiang, Xue Dayuan, eds., *Ziran baohu gailun* [An introduction to nature protection] (Beijing: Zhongguo huanjing kexue chuban she, 1991); Richard Louis Edmonds, *Patterns of China's Lost Harmony: A Survey of the Country's Environmental Degradation and Protection* (London: Routledge, 1994). See especially Judith Shapiro, *Mao's War Against Nature: Politics and the Environment in Revolutionary China* (Cambridge: Cambridge University Press, 2001), chapters 3 and 4 for a discussion of terracing unfertile hills to mimic Dazhai, filling in Dian Lake in the Kunming area, and reclamation of wetlands in northern China, etc.
4. Shapiro, *Mao's War Against Nature*, 57. Pan Wenshi, Director of Giant Panda Conservation and Research Center, interview by author, Beijing, PRC, January 2002.
5. The term "national treasure" is in quotes because this was an official term applied to the giant panda in 1972. The Chinese term is "国宝."

6. Wanglang ziran baohu chu diaocha zu, "Wanglang ziran baohu qu da xiongmao de chubu diaocha," (1974) 162.
7. Wanglang ziran baohu chu diaocha zu, "Wanglang ziran baohu qu da xiongmao de chubu diaocha," (1974) 162–173.
8. Based on the translation in Mao Zedong, "Talks at the Yenan Forum on Literature and Art," *Mao Tse-tung on Literature and Art* (Beijing: Foreign Languages Press, 1977 [1960]), 12.
9. Matthew David Johnson, "A Politics of Form: Cultural Conflict and China's Film Industry After the Great Leap Forward, 1959–1964," unpublished paper presented at the UCSD-Stanford Cultural Revolution Conference June 8–9, 2003, University of California, San Diego, 2, 18.
10. Johnson, "A Politics of Form," 2, 18.
11. Zhang Ding and Zhang Mei, "Shiyong meishu—Zhongying gongyi meishu xueyuan qizuo shixi" [Useful art—Work and practice of the Central Craft and Art Academy], *Renmin ribao* [People's Daily], June 4, 1961, 8.
12. I would like to thank Yun-Chiahn C. Sena for sharing her insights on brush painting. Clunas, *Art in China*, 144.
13. *Zhongguo youpiao quanji Zhongguo Renmin Gongheguo juan* (Beijing: Beijing Yanshan chuban she, 1989), 122.
14. "Chuantong gongyi kai xinhua" [Updating traditional art and craft], *Renmin ribao* [People's Daily], (June 5, 1973), 4.
15. "Hubei shou gongyi lao yiren qizuo yipi xin zuopin" [Old artists produce new art for Hubei handicrafts], *Renmin ribao* [People's Daily], August 27, 1972, 3.
16. Julia F. Andrews, *Painters and Politics in the People's Republic of China, 1949–1979* (Berkeley: University of California Press, 1994), 115–118.
17. "Jiji fazhan chuantong yigong meishu pin shenchan" (Enthusiastic development of traditional craft and artwork production). *Renmin ribao* (*People's Daily*), October 26, 1972, 2.
18. "Jiji fazhan chuantong gongyi," 2.
19. "Chuantong gongyi kai xinhua," 4.
20. "Shaoyang zhu yi kai xinhua" [Updating bamboo art in Shaoyang], *Renmin ribao* [People's Daily], June 23, 1973, 3.
21. "Shaoyang zhu yi kai xinhua," 3.
22. Hao Siyong, "Xiongmao wu zhou sui" ['Panda' is five years old], *Renmin ribao* [People's Daily], September 30, 1961, 5.
23. "Fazhan qing gongye yuanliao jidi Nei Menggu qing gongye bumen jianli jidi zhengqu yuanliao zhubu zi gei rui anbai hao ruping chang tongguo zhiyuan xumuye fazhan guang bi ru yuan" [The basic materials for the development of light industry. The departments involved in Inner Mongolia's light industry established basic materials to pursue the development of milk product production to support the development of husbandry and the expansion of milk resources], *Renmin ribao* [People's Daily], February 17, 1960, 2; "Nei Menggu dapi ru zhipin

gongying shichang" [A large supply of Inner Mongolian milk products hit the market], *Renmin ribao* [People's Daily], August 24, 1963, 2.

24. "Youqu de suliao wanju" [An interesting plastic toy], *Renmin ribao* [People's Daily], January 29, 1962, 2.
25. "Xin Xilan zongli Maerdeng fangwen Qinghua Daxue" [New Zealand Prime Minister Muldoon visits Qinghua University], *Renmin ribao* [People's Daily], April 30, 1976, 2.
26. "Xin Xilan zongli Maerdeng fangwen Qinghua Daxue" April 30, 1976, 2.
27. "Wenhua da geming tuidong zhaoxiangji gongye da fazhan: 1976 nian quanguo zhaoxiangji de chanliang, bi wenhua da geming qian de 1976 nian zengchangle 12 bei, pinzhong zengjia jin 4 bei. Shishi you li de pipan le Deng Xiaoping 'Jin buru xi'de miulun'" [The Great Proletarian Cultural Revolution promotes the development of the camera industry: in 1976 the production of cameras throughout the entire country was greater than before the Great Proletarian Cultural Revolution by twelve fold, the number of camera models increased by four fold, offering a genuine criticism of Deng Xiaoping's false assertion that 'Today is not as good as in the past'], *Renmin ribao* [People's Daily], June 14, 1976, 3.
28. "Wenhua da geming tuidong zhaoxiangji gongye da fazhan," 3.
29. "Wenhua da geming tuidong zhaoxiangji gongye da fazhan," 3.
30. "Nuli guanche zhixing Mao Zhuxi geming wenyi luxian de xin chengguo guoqing qijian zhuang shang ying yi pi xin yingpian" [Diligently carrying out Chairman Mao's line on art for national day with many new films], *Renmin ribao* [People's Daily], September 30, 1975, 4.
31. "Wo guo xinxing jiaopian gongye dale fanshen zhang" [Our nation brought about an upswing in the burgeoning film industry], *Renmin ribao* [People's Daily], February 12, 1976, 1.
32. "Wo keji yingpian zai Luoma fangying shoudao huanying" [Our technology film was well received in Rome], *Renmin ribao* [People's Daily], December 17, 1976, 6; "Wo yingpian zai Yindu guoji dianying jie shoudao huanying" [Our films were well received in India's international film festival], *Renmin ribao* [People's Daily], February 2, 1977, 5.
33. "Wo dui wai xie daibiao tuan jiesu dui Yilang youhao fangwen" [Our foreign relations envoy goes to Iran], *Renmin ribao* [People's Daily], September 11, 1977, 6; "Wo dui wai you xie daibiao tuan fangwen Yilake" [Our foreign relations envoy goes to Iraq], *Renmin ribao* [People's Daily], September 27, 1977, 5.
34. Zhong Zhaomin, "Zai Wanglang ziran baohu qu paishe kejiao yingpian, 'Xiongmao' gushipian, 'Xiongmao lixian yanzhi' deng yingpian de gaishu [Overview of the filming of the science education documentary, 'Panda' and the feature film 'Diary of a panda's adventures' in the Wanglang Nature Reserve]," personal reflections, April 2004.
35. *China Pictorial [Renmin huabao, 人民画报]*, May 1973, 22–25.

36. Naomi Oreskes, "Objectivity or Heroism? On the Invisibility of Women in Science," *Osiris*, Second Series, 11, *Science in the Field* (1996): 87–113.
37. Schneider, *Biology and Revolution in Twentieth-Century China*, 202–206.
38. Jiang Qing (wife of Mao Zedong), Wang Hongwen, Yao Wenyuan, and Zhang Chunqiao were four radical leaders of the Cultural Revolution who banded together and pushed the movement forward. After Mao's death in 1976, these four were designated as a clique, arrested, tried, and convicted, officially marking the end of the Cultural Revolution.

CHAPTER 5

1. "Beijing shi gewei hui zeng songgei meiguo ren min de yi dui xiongmao zai huashengdun guojia dongwuyuan juxing jiaojie yishi" [A ceremony was given in Washington, DC, at the National Zoo for the pair of pandas that the Beijing municipal revolutionary committee presented to the American people], *Renmin ribao* [People's Daily], April 22, 1972; "Dongjing juxing jieshou Zhongguo renmin zengsong yi dui da xiongmao yishi Erjie tang jin guanfang zhangguan, Qiaoben deng mei sanlang ganshi zhang, yi ji ge jie renshi canjia" [Tokyo held a reception for the pair of pandas given by the Chinese people; Commanding Officer, Head Secretary, and other public figures from other divisions all participated], *Renmin ribao* [People's Daily], November 5, 1972, 4.
2. Peter Enav, "Chinese Pandas Arrive in Taiwan in Charm Offensive," *Associated Press*, December 23, 2008.
3. Christina Lamb, "Peevish China recalls panda Tai Shan due to Obama's meeting with the Dalai Lama," *The Sunday Times* (UK), February 14, 2010.
4. Mark Magnier, "Attack of the Pandas," *Los Angeles Times*, March 21, 2006; Margaret MacMillan, *Nixon and Mao: The Week that Changed the World* (New York: Random House Trade Paperbacks, 2008), 148; "Youguan da xiongmao de lishi jizai" [A written account of the history of the giant panda], *Zhongguo ribao wangzhan* [China daily online], China Culture.org http://www.chinaculture.org/gb/cn_zggd/2006-01/11/content_77854.htm.
5. Kojima Noriyuki et al., eds., *Nihon Shoki* 3, (Tōkyō: Shōgakkan, 1994–1998), 218–219; Also, thank you to Kim Kono and Stephen Roddy for consultation on this passage and the contributors to the discussion of this myth at the Hatena Diary blog, http://d.hatena.ne.jp/syulan/20070523/p1 (Accessed March 4, 2008).
6. "Two Pandas are Presented to Bronx Zoo by Chinese at Ceremony in Chungking," *New York Times*, November 10, 1941, 19.
7. Zhong and Gipps, *2001 International Studbook for Giant Panda*, 1–2. Studbooks offer the pedigree of any given animal. This one lists all giant pandas captured and brought into captivity or born in captivity that survived long enough to receive a number, and usually a name.
8. Xie and Gipps, *2001 International Studbook for Giant Panda*, 2.

9. Xie and Gipps, *2001 International Studbook for Giant Panda*, 7, 10, 11.
10. "Peng Zhen shizhang zenggei Pingrong yi pi zhengui dongwu" [Mayor Peng Zhen gave Pyongyong several precious animals], *Renmin ribao* [People's Daily], June 8, 1965, 4.
11. Xie and Gipps, *2001 International Studbook for Giant Panda*, 7, 10, 11.
12. Zhang Xiruo daibiao de fayan [Representative Zhang Xiruo's speech], "Huxiang xiqu, huxiang fazhan, huxiang zunzhong fazhan dui wai wenhua jiaoliu gongzuo" [Mutual assimilation, mutual development, mutual respect, development of cultural foreign exchange work], published in *Renmin ribao* [People's Daily], May 4, 1959, 5.
13. Joseph A. Davis, "A China Doll and Pal Come to Town," *New York Times*, April 23, 1972, E5.
14. "Nikesong zongtong furen zai jing canguan youlan" [President Nixon's wife on a sightseeing tour of the capital], *Renmin ribao* [People's Daily], February 23, 1972, 2.
15. "Beijing shi gewei hui daizhuren Wu De huijian husong sheniu de meiguo keren" [Beijing municipal revolutionary committee representative Wu De meets and escorts musk oxen guests from America], *Renmin ribao* [People's Daily], April 13, 1972, 5; "Beijing shi gewei hui zeng songgei meiguo ren min de yi dui xiongmao," *Renmin ribao* [People's Daily], April 22, 1972, 6; "The People's Pandas" *New York Times*, March 15, 1972, 46.
16. "Beijing shi gewei hui daizhuren Wu De," April 13, 1972.
17. "Beijing shi gewei hui daizhuren Wu De," April 13, 1972.
18. "Security is Tight as 2 Pandas Land, Gifts from Peking Housed at Zoo in Washington," *New York Times*, April 17, 1972, 5; *New York Times*, April 15, 1972, 62.
19. Nan Robertson, "New Pandas Melt Hearts at National Zoo," *New York Times*, April 18, 1972, 1.
20. "Beijing shi gewei hui zeng songgei meiguo ren min de yi dui xiongmao," April 22, 1972, 6. Their popularity only grew throughout their lives in the National Zoo of Washington, DC, and the city mourned at their deaths.
21. "Panda-monium in Washington," *New York Times*, June 11, 1972.
22. "People's Pandas," *New York Times*, March 15, 1972, 46.
23. "People's Pandas," *New York Times*, March 15, 1972, 46.
24. Robertson, "New Pandas," April 18, 1972, 1.
25. Michael Gross, Cover Art, *National Lampoon*, (July 1972).
26. Robertson, "New Pandas," April 18, 1972, 1.
27. Harrison E. Salisbury, "Dinner with Mrs. Sun Yat-sen in Old Peking," *New York Times*, June 3, 1972, 2.
28. Harrison E. Salisbury, "U.S. Musk Oxen Recuperating, Draw Crowds in Peking Zoo," *New York Times*, June 14, 1972, 2.
29. Samuel S. Kim, "The People's Republic of China in the United Nations: A Preliminary Analysis," *World Politics* 26, no. 3 (April 1974), 301; Justin S.

Hempson-Jones, "The Evolution of China's Engagement with International Governmental Organizations: Toward a Liberal Foreign Policy," *Asian Survey* 45, no. 5 (2005), 707.

30. Another comparative survey of *Renmin ribao* articles must be done to evaluate this relationship properly. My pool is based on articles that mention "panda" or "giant panda" in the text. I would be surprised if there was a great discrepancy between the ratio of articles about China's new relations with these countries that do include "panda" or "giant panda" and those that do not; however, this caveat must be expressed until a more exact assessment is made.
31. "Dongjing juxing jieshou Zhongguo renmin zengsong yi dui da xiongmao," November 5, 1972, 4; "China Presents Japan with Two Giant Pandas," *New York Times*, September 29, 1972.
32. *New York Times*, October 29, 1972, 17.
33. "Dongjing juxing jieshou da xiongmao," November 5, 1972, 4.
34. Masaya Tsuchiya, "Recent Developments in Sino-Japanese Trade," *Law and Contemporary Problems* 38, no.2, Trade with China (Summer-Autumn, 1973): 241–243.
35. Chalmers Johnson, "The Patterns of Japanese Relations with China, 1952–1982," *Pacific Affairs* 59, no. 3 (Autumn, 1986): 404.
36. Chalmers Johnson, "Patterns of Japanese Relations with China," 403.
37. The Nixon administration not only had declined to inform Japan about this important policy change, it also did not inform the US public, Congress, the State Department, or even the Secretary of State that this was going to happen. Henry Kissinger, the National Security Advisor, made these arrangements with Zhou Enlai only shortly before Nixon announced it publicly. See Chalmers Johnson, "Patterns of Japanese Relations with China," 410.
38. Chalmers Johnson, "Patterns of Japanese Relations with China," 411.
39. "Dongjing juxing jieshou da xiongmao," November 5, 1972, 4.
40. Lee W. Farnsworth, "Japan 1972: New Faces and New Friends," *Asian Survey* 13, no. 1 (January, 1973): 122.
41. Farnsworth, "Japan 1972: New Faces and New Friends," 113.
42. "Hu yin shanlu gong youyi—ji di shi ci ri zhong youhao xingnian xialing ying" [Home in the hidden foothills praising friendship—records of the tenth Japanese-Chinese friendship summer camp for young people], *Renmin ribao* [People's Daily], August 8, 1973, 5.
43. "Wo zengsong xiongmao jiaojie yishi zai pali juxing Ceng Tao dashi he Fengdanei bu zhang gong zhu Zhong Fa youyi bu duan fazhan" [At the ceremony for our gift of pandas, Ambassador Ceng Tao and Ministry Head Fengdanei together celebrated the continuous development of friendship between China and France], *Renmin ribao* [People's Daily], December 22, 1973, 5.
44. "Wo guo zhengfu zengsong Pengpishe zongtong he Faguo renmin de yi dui da xiongmao yunwang Pali" [The pair of pandas that our national government gave

President Pompidou and the people of France are headed for Paris], *Renmin ribao* [People's Daily], December 9, 1973, 4.

45. "Wo zeng Yingguo renmin da xiongmao jioajie yishi zai lundun juxing" [The handover ceremony for the giant pandas we gave to the English people was held in London], *Renmin ribao* [People's Daily], November 8, 1974, 6.
46. "Wo zengsong de yidui daxiongmao yun di Moxige" [Our gift pandas have arrived in Mexico], *Renmin ribao* [People's Daily], September 12, 1975, 5.
47. Huang Hua wei wo ping pang qiu daibiao tuan juxing shaodai hui dui Meiguoren he pingxie deng gei yu daibiaotuan reqing huanying he youyu gao jiedai biaoshi ganxie" [Huang Hua acts as our Ping-Pong team representative and holds a reception to express gratitude to the American people and the Ping-Pong association for giving our team a warm welcome and friendship], *Renmin ribao* [People's Daily], April 23, 1972, 6.
48. "Huang Hua wei wo ping pang qiu daibiao," April 23, 1972, 6.
49. "Ying hua shi jie fang dong ling" [Visiting eastern neighbors during the cherry blossom festival], *Renmin ribao* [People's Daily], May 10, 1973, 5.
50. "Tanaka (Tianzhong jiao rong) Shouxiang huijian Liao Chengzhe tuanzhang Riben xinwen jie pengyou juxing cha hui huanying wo daobiao tuan" [Prime Minister Tanaka Kakuei met committee head Liao Chengzhi; Japan's news friends hosted a tea ceremony to welcome our representative committee], *Renmin ribao* [People's Daily], May 15, 1973, 5.
51. "Zhong Ri youhao de xin bianzhang—ji Zhong Ri youhao xiehui daibiao tuan fangwen Riben" [New chapters in Chinese-Japanese friendship—records of the Chinese-Japanese friendship association representatives' visit to Japan], *Renmin ribao* [People's Daily], June 4, 1973, 5.
52. Hu Tieqing, interview by author, January 14, 2002.
53. Xie and Gipps, *2001 International Studbook for Giant Panda*, 2–11.
54. Xie and Gipps, *2001 International Studbook for Giant Panda*, 2–11.
55. Xie and Gipps, *2001 International Studbook for Giant Panda*, 11–23.
56. Dong Zhiyong, ed., *Zhongguo linye nianjian*, 80.
57. "Beijing shi gewei hui zeng songgei meiguo ren min de yi dui xiongmao," April 22, 1972, 6; "Dongjing juxing jieshou da xiongmao," November 5, 1972, 4.
58. Xie and Gipps, *2001 International Studbook for Giant Panda*, 2–23.
59. Xie and Gipps, *2001 International Studbook for Giant Panda*, 11–23.
60. Han Zhengfu, "Han Zhengfu tongzhi zai quan sheng zhengui dongwu ziyuan, diaocha zuotan huiyi shang de jianghua" [Comrade Han Zhengfu's speech at the province-wide meeting on precious animal resources and surveys], November 17, 1973, 3.
61. PRC, Nonglin bu [Ministry of Agriculture and Forestry], "Guanyu zheng qiu dui 'yesheng dongwu ziyuan baohu tiaoli' (caoan) de yijian he zanting pizhuo buzu guojia zhengui dongwu de tongzhi" [Report concerning the government's request about the suggestion to temporarily halt the approval to trap precious

species with the 'wild animal resource protection regulations' (draft)], (3 Nonglin [lin] zi di 53 hao), May 8, 1973, 1.

62. Xie and Gipps, *2001 International Studbook for Giant Panda*, 23–51.
63. Han Zhengfu, "Zai quan sheng zhengui dongwu ziyuan," November 17, 1973.
64. Han Zhengfu, "Zai quan sheng zhengui dongwu ziyuan," November 17, 1973.
65. Sichuan sheng, Pingwu xian, Linye ju [Sichuan Province, Pingwu County, forestry bureau], "Guanyu caoni 'yesheng dongwu ziyuan baohu tiaojian' de shuoming" [Draft of explanation of 'factors in terms of wild animal resource protection'], May 1973; Han Zhengfu, "Zai quan sheng zhengui dongwu ziyuan," November 17, 1973; PRC, Nonglin bu (3 Nonglin [lin] zi di 53 hao), May 8, 1973.
66. PRC, Nonglin bu (3 Nonglin [lin] zi di 53 hao), May 8, 1973.
67. Han Zhengfu, "Zai quan sheng zhengui dongwu ziyuan," November 17, 1973, 3, 5.
68. PRC, Nonglin bu (3 Nonglin [lin] zi di 53 hao), May 8, 1973, 3.
69. Paul G. Pickowicz, Professor of History, University of California, San Diego, interview by author May 6, 2004, La Jolla, California.

CHAPTER 6

1. Some of the material used in this chapter is incorporated into the co-authored article, Sigrid Schmalzer and E. Elena Songster, "Wild Pandas Wild People: Two Views of Wilderness in Deng-Era China," in *Visualizing Modern China: Image, History, and Memory, 1750–Present*, ed. James A. Cook, Joshua Goldstein, Matthew D. Johnson, and Sigrid Schmalzer. (Lanham, MD: Lexington Books, 2014), 259–278; Sichuan sheng, Pingwu xian, Geming weiyuanhui [Sichuan Province, Pingwu County, Revolutionary committee], "Guanyu baohu da xiongmao de jiji tongzhi" [Urgent notice concerning protecting giant pandas] (Pingge fa 76 di 12 hao), February 10, 1976, 1; PRC, Nonglin bu [Ministry of Agriculture and Forestry], "Guanyu jiaqiang da xiongmao baohu gongzuo de jiji tongzhi" [Urgent notice concerning giant panda protection work], (Nonglin 76 [lin] zi di 20 hao), March 16, 1976, 1; Sichuan sheng, Linye ting [Sichuan province, Forestry department], "Guanyu guanche zhixing nonglin bu 'guanyu jiaqiang da xiongmao baohu gongzuo de jiji tongzhi' de tongzhi" [Notice regarding carrying out the policy of the Ministry of Agriculture and Forestry 'Urgent notice regarding strengthening giant panda protection work'], (Chuanlin zao 76 di 27 hao), March 23, 1976, 1; Sichuan sheng, Linye ting [Sichuan Province, Forestry department], "Qing shouji da xiongmao ziran siwang qingkuang de jiji tongzhi" [Urgent notice requesting the collection of giant pandas that naturally died], (Chuanlin zao 76 di 29 hao), April 9, 1976, 1; Sichuan sheng, Pingwu xian, Geming weiyuanhui [Sichuan Province, Pingwu County, Revolutionary Committee], "Guanyu diaocha da xiongmao qingkuang de jiji tongzhi" [Urgent notice regarding a survey of the giant panda situation], (Pingge fa 76 di 25 hao),

April 13, 1976, 1; Da xiongmao ziran siwang lianhe diaocha dui [Giant panda natural death United Survey Team (hereafter United Survey Team)], July 20, 1976, 1; Sichuan sheng, Mianyang, Linye ju [Mianyang prefecture forestry bureau], "Guanyu jiaqiang dui da maoxiong xiankuang guancha he dui jianzhu kaihua siwang hou de huifu qingkuang jinxing diaocha zongjie de tongzhi" [Summary notice regarding the strengthening of the monitoring of the present giant panda situation and the survey of the advancement of post-bamboo flowering and die-off recovery], (Dilin jingying di 32 hao), March 13, 1976.

2. Sichuan sheng, Mianyang linye ju [Mianyang Forestry Bureau], (Dilin jingying 76 di 32 hao), March 13, 1976, 1.
3. Yang Ruoli, Zhang Fuyun, and Luo Wenying, "1976 nian da xiongmao zainan xing siwang yuanyin de shenlun" [Probing into reasons for the 1976 catastrophic death of giant pandas], *Acta Theriologica Sinica* 1, no. 2 (December 1981): 128.
4. Jiang Tingan, "Zai da xiongmao de guxiang" [In panda country], *Bowu* [Natural history] 3 (November 1980): 15. The area that this article describes is most likely Wanglang and Pingwu County. This identification is indicated by the fact that at that time there were still very few nature reserves and the type of bamboo described, arrow bamboo, was the predominant type of bamboo that grew in Wanglang. While there are Tibetans living in Pingwu County, there are more local Baima. Because of their official categorization as Baima Tibetans, many official and popular documents refer to them simply as Tibetan. This is,therefore, most likely a reference to a member of the Baima group rather than a Tibetan.
5. Pingwu County is in northern Sichuan and neighbors southern Gansu. Both areas contain pandas and are situated along the same Min mountain range.
6. Yang et al., "1976 da xiongmao siwang yuanyin," 128.
7. Sichuan sheng, Mianyang linye ju (Dilin jingying 76 di 32 hao), March 13, 1976.
8. For a brief overview of the major events during this period, please see Jonathan D. Spence, *The Search for Modern China* (New York: W. W. Norton & Co., 2013 [1990]), 559–638.
9. Yao Weimin, ed., *Zhonggong Pingwu difang shi dashi ji, 1935–1998* [Central Pingwu local history and record of major events, 1935–1998], (Pingwu: Sichuan donghua yingwu jituan youxian gongsi, 1999), 151.
10. PRC, Nonglin bu [Ministry of Agriculture and Forestry], (Nonglin 76 [lin] zi di 20 hao), March 16, 1976; Sichuan sheng, Linye ting [Sichuan Province, Forestry Department], (Chuanlin zao 76 di 29 hao), April 9, 1976; Sichuan sheng, Pingwu xian, Geming weiyuanhui, (Pingge fa 76 di 12 hao), February 10, 1976; Sichuan sheng, Pingwu xian, Geming weiyuanhui, (Pingge fa 76 di 25 hao), April 13, 1976; Sichuan sheng, Pingwu xian, Geming weiyuanhui, (Pingge fa 76 di 29 hao), April 22, 1976.
11. While it is believed that environmental factors can influence a bamboo's flowering cycle, the main trigger, although not fully understood, is considered to be

independent of the environment. Zhong Zhaomin, "Zhulei kaihua yu dizhen de guanxi," undated and unpublished essay, 2.

12. Sichuan sheng, Pingwu xian, Geming weiyuanhui, (Pingge fa 76 di 12 hao), February 10, 1976, 1.
13. Sichuan sheng, Pingwu xian, Geming weiyuanhui, (Pingge fa 76 di 12 hao), February 10, 1976, 1–2.
14. Sichuan sheng, Pingwu xian, Geming weiyuanhui, (Pingge fa 76 di 12 hao), February 10, 1976, 1.
15. Yang et al., "1976 da xiongmao siwang yuanyin," 128.
16. Guo An, "Wo guo zuida de dongwu yuan," May 6, 1956, 3; Zhou Jianren, "Guanyu xiongmao," July 6, 1956, 8.
17. Sichuan sheng, Pingwu xian, Geming weiyuanhui, (Pingge fa 76 di 12 hao), February 10, 1976, 2.
18. Sichuan sheng, Linye ting, (Chuanlin zao 76 di 27 hao), March 23, 1976, 2.
19. Sichuan sheng, Linye ting, (Chuanlin zao 76 di 29 hao), April 9, 1976, 1–2.
20. PRC, Nonglin bu, (Nonglin 76 [lin] zi di 20 hao), March 16, 1976, 2.
21. Sichuan sheng, Pingwu xian, Geming weiyuanhui, (Pingge fa 76 di 25 hao), April 13, 1976, 5.
22. Sichuan sheng, Pingwu xian, Geming weiyuanhui, (Pingge fa 76 di 29 hao), April 22, 1976, 1.
23. Sichuan sheng, Pingwu xian, Geming weiyuanhui, (Pingge fa 76 di 12 hao), February 10, 1976, 2.
24. Sichuan sheng, Pingwu xian, Geming weiyuanhui, (Pingge fa 76 di 25 hao), April 13, 1976, 6.
25. Schaller et al., *Giant Pandas of Wolong*, 254.
26. Gazetteers are local histories. They commonly compile centuries of history and include details about local flora, fauna, and natural disasters among many other types of information including local agriculture, industry, population. Sometimes they are written at the district level, other times the provincial level; there are also topical national-level gazeteers.
27. Sichuan sheng, Mianyang, Linye ju, (Dilin jingying 76 di 32 hao), March 13, 1976, 1–2.
28. PRC, Guowu yuan "Guowu yuan guanyu jiji baohu he heli liyong yesheng dongwu zi yuan de zhishi," (Guolin Tan zi 287 hao), September 14, 1962, 5; PRC, Nonglin bu, "Guanyu jiaing da xiongmao baohu," (Nonglin 76 [lin] zi di 20 hao), March 16, 1976, 1.
29. PRC, Nonglin bu, "Guanyu jiaing da xiongmao baohu," (Nonglin 76 [lin] zi di 20 hao), March 16, 1976, 2.
30. Sichuan sheng, Linye ting, (Chuanlin zao 76 di 29 hao), April 9, 1976, 2.
31. Qian Danning, ed., *Pingwu Xianzhi*, 505.
32. Qian Danning, *Pingwu xianzhi*, 506.
33. Sichuan sheng, Pingwu xian, Linye ju [Sichuan Province, Pingwu County, Forestry Bureau], "Guanyu jiaqiang Wanglang ziran baohu qu guanli gongzuo

de baogao" [Report concerning the strengthening of Wanglang Nature Reserve management work], (Pinglin hu 76 zi di 03 hao), May 20, 1976, 1.

34. Sichuan sheng, Pingwu xian, Linye ju [Sichuan Province, Pingwu County, Forestry Bureau], (Pinglin hu 76 zi di 03 hao), May 20, 1976, 2. The other panda destined for Paris was captured in Baoxing County in southern Sichuan. Xie and Gipps, *2001 International Studbook for Giant Panda*, 14.
35. Sichuan sheng, Pingwu xian, Linye ju [Sichuan Province, Pingwu County, Forestry Bureau], "Guanyu Wanglang ziran baohu qu 76 nian jijian kuan kai ji jihua baogao" [Report concerning Wanglang Nature Reserve construction budget for 1976], (Pinglin ban76 zidi 04 hao), June 18, 1976.
36. One *jin* is equal to 1.1023 lbs.
37. Sichuan sheng, Pingwu xian, Linye ju [Sichuan Province, Pingwu County, Forestry Bureau], "Guanyu qingshi jiejue si da xiongmao shiliang gongying de baogao" [Report concerning request to solve the problem with supplies for feeding and raising giant pandas], (Pinglin ban 76 zi di 06 hao), June 22, 1976, l.
38. Sichuan sheng, Pingwu xian, Geming weiyuanhui [Sichuan Province, Pingwu County, Revolutionary Committee], "Guanyu zhuan fa Sichuan sheng geming weiyuanhui yingfa nonglin bu song 'guanyu da xiongmao daliang siwang qingkuang de diaocha baogao' de han" [Reply regarding the forwarding of Sichuan Province's Revolutionary Committee's issuing of the Ministry of Agriculture and Forestry's directive, 'Concerning the survey report of the large number of giant panda deaths'], (Pingge fa 76 di 78 hao), December 14, 1976, 1; Da xiongmao ziran siwang lianhe diaocha dui, United Survey Team, July 20, 1976, 1; Zhong Zhaomin, "Da xiongmao siwang yu dizhen de guanxi" [The relationship between giant panda death and earthquakes], undated essay, 1; Wen Zhe and Wang Menghu, "Da xiongmao yu zhu" [Giant pandas and bamboo], *Da ziran* [Nature] 1 (1980): 12; Jiang Tingan, "Zai da xiongmao de guxiang," 15; Schaller et al., *Giant Pandas of Wolong*, 255.
39. Jiang Tingan, "Zai da xiongmao de guxiang," 15–16; Yang et al., "1976 da xiongmao siwang yuanyin," Zhong Zhaomin, "Da xiongmao siwang yu dizhen," 1.
40. As a reminder, studbooks offer the pedigree of any given animal. This one in particular lists all giant pandas captured and brought into captivity or born in captivity that survived long enough to receive a number, and also usually a name up to 2001. Xie and Gipps, *2001 International Studbook for Giant Panda*, 18–20.
41. They would have died quickly, if this was the case, because even unnamed "New No. 7" was recorded in the captive-panda studbook after only a three-month life in captivity before she died. Xie and Gipps, *2001 International Studbook for Giant Panda*, 18.
42. Stephen J. O'Brien, Pan Wenshi, and Lü Zhi, "Pandas, People and Policy," *Nature* 369 (May 19, 1994): 180.
43. Schaller, *Last Panda*, 200.

44. Deduced from population parameters stated in Schaller, *Last Panda*, 230; Christopher S. Wren, "Bureaucracy and Blight Imperil China's Pandas," *New York Times*, July 3, 1984, C1.
45. Schaller, *Last Panda*, 200; Alan H. Taylor and Qin Zisheng, "Culm Dynamics and Dry Matter Production of Bamboos in the Wolong and Tangjiahe Giant Panda Reserves, Sichuan, China," *Journal of Applied Ecology* 24, no. 2 (August 1987): 422; Donald G. Reid and Hu Jinchu, "Giant Panda Selection Between Bashania Fangiana Bamboo Habitats in Wolong Reserve, Sichuan, China," *Journal of Applied Ecology* 28 (1991): 230–231; Schaller et al., *The Giant Pandas of Wolong*, 255.
46. Schaller et al., *The Giant Pandas of Wolong*, 255.
47. Yin Hong, "Zhongguo yesheng dongwu baohu xiehui fu zeren tan muqian da xiongmao de zaiqing ji jixu de jiuzai cuoshi [Head of China's Wild Animal Protection Association discusses the present situation of the giant panda crisis and the urgent need for panda rescue measures]," *Renmin ribao* [People's Daily], April 16, 1984, 3; Schaller, *Last Panda*, 201.
48. RMRB, January 1975–January 1985.
49. RMRB, August 21, 1983, 3.
50. RMRB, September 1, 1983, 2.
51. Yin Hong, "Zhongguo yesheng dongwu baohu xiehui fuze ren tan," 3.
52. Dong Zhiyong, "Qiangjiu 'guobao' da xiongmao" [Saving the 'national treasure' the giant panda], interview by journal reporter in, *Yesheng dongwu* [Chinese Wildlife] 3 (May 1984): 1.
53. RMRB, September 1, 1983.
54. Dong Zhiyong, "Qiangjiu 'guobao' da xiongmao," 2.
55. Enid Nemy, "First Lady's Quest: 'The Real China,'" *New York Times*, April 23, 1984, A6.
56. *New York Times* (hereafter NYT), January 22, 1984, 34; NYT, February 5, 1984, 28.
57. RMRB, August 17, 1983, 1.
58. Yin Hong, "Zhongguo yesheng dongwu baohu xiehui fuze ren tan," 3; Dong Zhiyong, "Qiangjiu 'guobao' da xiongmao," 2.
59. Hu Jinchu, "Canjia Zhongguo dongwuxue hui chengli wushi zhou nian nian hui de dongwuxue gongzuozhe huyu: caiqu jinji jieshi baohu da xiongmao" [Participating in the fiftieth annual meeting of China's national zoological meeting, a zoological worker's appeal: take urgent measures to protect the giant panda], *Yesheng dongwu* [Chinese wildlife] 5 (September 1984): 2.
60. Hu Jinchu, "Canjia Zhongguo dongwuxue hui," 2.
61. Dong Zhiyong, "Qiangjiu 'guobao' da xiongmao," 2.
62. Ouyang Huiyun, "Wolong ziran baohu qu da xiongmao zhuan yi dao anquan didai" [Moving giant pandas in the Wolong Nature Reserve to a safer area] *Renmin ribao* [People's Daily], October 22, 1983, 3; Yin Hong, "Zhongguo yesheng dongwu baohu xiehui fuze ren tan," 3; Dong Zhiyong, "Qiangjiu 'guobao' da xiongmao," 2.

63. Schaller, *Last Panda*, 202.
64. Dong Zhiyong, "Qiangjiu 'guobao' da xiongmao," 1.
65. Hu Jinchu, "Canjia Zhongguo dongwuxue hui," 1.
66. Chen Yuanfei, "Da xiongmao 1983 dashi ji" [Major giant panda events in 1983], *Da ziran tansuo* [Exploring Nature] 2 (1984): 185.
67. This may simply be the result of inaccurate reporting, poor information, or pressure on the journalists to produce dead pandas for public consumption. Regardless, George Schaller was alarmed by reading news in the United States claiming that pandas had died in Wolong. When he checked this information with researchers in Wolong, they asserted that it was incorrect and that no pandas in Wolong had died of starvation. Schaller, *Last Panda*, 202.
68. Ouyang Huiyun, "Wolong ziran baohu qu da xiongmao zhuan yi," 3.
69. *New York Times*, February 13, 1984, A6.
70. Kenneth G. Johnson, George B. Schaller, and Hu Jinchu, "Response of Giant Pandas to a Bamboo Die-off," *National Geographic Research* 4 (1988): 161–177.
71. Dong Zhiyong, "Qiangjiu 'guobao' da xiongmao," 1.
72. Yin Hong, "Zhongguo yesheng dongwu baohu xiehui fuze ren tan," 3.
73. Hu Jinchu, "Canjia Zhongguo dongwuxue hui," 1.
74. RMRB, May 17, 1984, 3.
75. Wren, "Bureaucracy and Blight Imperil China's Pandas," C1.
76. Qian Danning, *Pingwu xianzhi*, 421. Although Pingwu is a different county, the reporter for the *New York Times* indicated that the sum of the reward was approximately one-year's salary. The average yearly salary for a Pingwu resident in 1984 was 486 *yuan* per year. It is more likely that the sum was mistranslated into a dollar amount than that the standard of living in another mountainous county of Sichuan was eight times higher than that of Pingwu County.
77. Dong Zhiyong, "Qiangjiu 'guobao' da xiongmao," 2.
78. Dong Zhiyong, "Qiangjiu 'guobao' da xiongmao," 2.
79. Liu Xiangji and Huang Zhenggen, "Baohu da xiongmao deng zhenxi dongwu yesheng dongwu baohu xiehui kaizhan mujuan huodong" [Protecting the giant panda and other precious and rare animals, the Wild Animal Protection Association develops fundraising activity], *Renmin ribao* [People's Daily], January 9, 1984, 3.
80. RMRB, April 4, 1984, 3.
81. RMRB, June 8, 1984, 6.
82. RMRB, February 22, 1984, 3.
83. Yin Hong, "Zhongguo yesheng dongwu baohu xiehui fuze ren tan," 3.
84. *New York Times*, January 22, 1984, 34.
85. *New York Times*, April 18, 1984, A24.
86. Enid Nemy, "At End of Day for Reagans, Dinner with Nine Courses," *New York Times*, April 28, 1984, 4.
87. RMRB, April 29, 1984, 1; Dong Zhiyong, "Qiangjiu 'guobao' da xiongmao," 2.

88. RMRB, March 16, 1984, 1.
89. RMRB, February 24, 1984, 3.
90. Liu Yuanyun, "Chengdu lujun xuexiao wei qiangjiu da xiongmao yongyue juankuan juanliang" [The Chengdu army academy enthusiastically contributed money and grain to save the giant panda], *Sichuan dongwu* [Sichuan animals] 3, no. 3 (August 1984): 46.
91. Xi Zefu, "Chengdu shi di shi zhongxue shi sheng jiji juankuan zhengjiu da xiongmao" [Teachers and students and Chengdu's Number Ten middle school enthusiastically contributed funds to save the giant panda], *Sichuan dongwu* [Sichuan Animals] 3, no. 3 (August 1984): 46
92. Dong Zhiyong, "Qiangjiu 'guobao' da xiongmao," 2.
93. Hu Jinchu, "Canjia Zhongguo dongwuxue hui," 3.
94. Song Lianfeng, "Wolong leng jian zhu mianji kaihua gusi da xiongmao wunai xia shan jieshi huagu zhu" [Because the arrow bamboo has mass-flowered and died in Wolong, giant pandas have no choice but to descend to eat the umbrella bamboo], *Renmin ribao* [People's Daily], April 22, 1984, 1.
95. Wren, "Bureaucracy and Blight Imperil China's Pandas," C1.
96. Wren, "Bureaucracy and Blight Imperil China's Pandas," C1; Christopher S. Wren, "Chinese Official Denies Gift for Pandas was Sidetracked," *New York Times*, October 17, 1984, A19.
97. Song Houqing, "Linye bu fu buzhang Dong Zhiyong shuo: qiangjiu da xiongmao shi xiang chang qi gongzuo" [The vice-chair of the Ministry of Forestry, Dong Zhiyong said: saving the giant panda is a long-term task], *Renmin ribao* [People's Daily], December 18, 1984, 3.
98. Xie and Gipps, *2001 International Studbook for Giant Panda*, 27–30.
99. Song Houqing, "Dong Zhiyong shuo: qiangjiu da xiongmao shi xiang chang qi gongzuo," 3.
100. Johnson et al., "Responses of Giant Pandas to a Bamboo Die-off," 176.
101. Johnson et al., "Responses of Giant Pandas to a Bamboo Die-off," 161–177.
102. Johnson et al., "Responses of Giant Pandas to a Bamboo Die-off," 176.
103. Johnson et al., "Responses of Giant Pandas to a Bamboo Die-off," 176.
104. Pan et al., *Jiuxiu shengcun de jihui*, 3.
105. Pan et al., *Jiuxiu shengcun de jihui*, 7.
106. Pan et al., *Jiuxiu shengcun de jihui*, 5.
107. Pan Wenshi, professor of Biology at Peking University, interview by author, 2001, Beijing, PRC.
108. RMRB, January 5, 1985– November 29, 1987.

CHAPTER 7

1. KarmaQuest, ecotourism travel website. Content available from http://www.karmaquests.com/sichuan-2003.htm (content confirmed 21 March and 25

June 2004). Content has changed and the new content is available at http://www.karmaquests.com/sichuan_panda_trip.htm (Accessed February 5, 2017, website updated 2014).

2. Geoff Carey, ed., *A Biodiversity Review of China* (Hong Kong: World Wide Fund for Nature [WWF] International, 1996), 4. James Harkness, interview by author, September 2001; Lü Zhi, correspondence, August 13, 2015.
3. Li Shengzhi, ICDP representative. Interview by author, May 25, 2002, Pingwu City, Pingwu County, Sichuan.
4. Integrated Conservation Development Program (ICDP), June 1998, 53, (hereafter, ICDP),
5. ICDP, 48.
6. Pingwu xian, linye ju, Pingwu xian shoulie gongzuo kaizhan qingkuang jianjie [A concise report on the status of the development of work on hunting in Pingwu County] (September 13, 1965), 6.
7. Immanuel C.Y. Hsü, *China Without Mao: The Search for a New Order*, 2nd ed. (New York: Oxford University Press, 1990), 188.
8. "Outline of the Protection Project for the Giant Panda and its Habitat in Sichuan Province," List of Proposals, submitted to WWF, 1993.
9. Carey, ed., *A Biodiversity Review of China*, 4.
10. James Harkness, interview, 2001; Lü Zhi, correspondence, 2015.
11. Lü Zhi, Executive Director of the Peking University Center for Nature and Society and founder of the Shanshui Conservation Center College of Life Science, Peking University. Interview by author, July 10, 2002.
12. Lü Zhi, interview, July 10, 2002.
13. Chen Youping, Director of Wanglang Nature Reserve, interview by author May 28, 2002, Pingwu County, Sichuan, PRC.
14. Baima community members, interview with author, 2001, 2005.
15. Jiang Shiwei, Deputy Director of Wanglang Nature Reserve, interview by author, May 29, 2002, Pingwu County, Sichuan, PRC.
16. ICDP, 55.
17. Taylor and Qin, "Culm Dynamics and Dry Matter Production of Bamboos," 419–433.
18. Ronald R. Swaisgood, Zejun Zhang, Fuwen Wei, David E. Wildt, and Andrew J. Kouba, "Giant Panda Conservation Science: How Far We Have Come," *Biology Letters* 6.2 (2010): 143–145.
19. Pan Wenshi et al., *Jixu shengcun de jihui*, 61–62; Wang Dajun, "Postscript: The People of the Qinling Study," in Pan Wenshi et al., *A Chance for Lasting Survival: Ecology and Behavior of Wild Giant Pandas*, trans. Richard B. Harris (Washington, D.C.: Smithsonian Institution Scholarly Press, 2014), 329.
20. Qian Danning ed., *Pingwu xianzhi*, 847.
21. Zhang Shougong, "China's 1998 Flood Disaster, Cause and Response," Presented at conference, "Natural Disaster and Policy Response in Asia, Implications for Food Security," Harvard University Asia Center, Spring 1999.

22. Lü Zhi, interview, July 10, 2002.
23. Wanglang National Nature Reserve website: http://www.wanglang.com/Display.asp?ID=26 (Accessed August 12, 2010).
24. TIES, The International Ecotourism Society, "TIES Global Ecotourism Factsheet," TIES, September, 2006, www.ecotourism.org (Accessed July 7, 2011).
25. Karmaquest, http://www.karmaquests.com/index.htm (Accessed August 10, 2010).
26. Karmaquest, http://www.karmaquests.com/about-us.htm (Accessed August 12, 2010).
27. Chen Youping, interview, May 28, 2002.
28. Li Shengzhi, interview, October 2001.
29. Kang Jia (pseudonym), Baima woman, interview by author, Pingwu County, Sichuan, PRC, May 29, 2002.
30. World Wide Fund for Nature (WWF), China. http://www.wwfchina.org/english/loca.php?loca=107#2. Accessed December 18, 2012.
31. Emily Yeh and Chris Coggins, eds., *Mapping Shangrila: Contested Landscapes in the Sino-Tibetan Borderlands* (Seattle: University of Washington Press, 2014), especially 95–197.
32. Zhang, Zhenguo et al., "Ecotourism and Nature-Reserve Sustainability in Environmentally Fragile Poor Areas: The Case of the Ordos Relict Gull Reserve in China," *Sustainability: Science, Practice, & Policy* 4, No. 2 (Fall/Winter 2008), 13–14. http://ejournalnbii.org (Accessed August 8, 2010).
33. "Sichuan Wanglang, Tangjiahe ziran baohuqu xuexi kaocha baogao [Sichuan, Wanglang, Tangjiahe nature reserve study and survey report]," Mianyang shi linyeju zhuban [Mianyang City Forestry Office], December, 30, 2005.
34. Jiang Shiwei, "Wanglang: Zai minshan shenchu jingjing zhanfang [Wanglang: the quiet blossom deep in the heart of the Min Mountains]," *Zhongguo luse shibao China's Green Times*, August 6, 2010. http://www.forestery.gove.cn/portal/main's'72/content-434604.html (Accessed August 7, 2010).
35. Guangming He, Xiaodong Chen, Wei Liu, Scott Bearer, and Shiqiang Zhou, "Distribution of Economic Benefits from Ecotourism: A Case Study of Wolong Nature Reserve for Giant Pandas in China," *Environmental Management* 42 (2008): 1021.
36. He et al. (2008), 1024.
37. Pan Wenshi, interviews by author, February 2002.
38. Chen Liang, "Ecotourism to Save Nature," *China Daily*, November 14, 2003, http://www.chinadaily.com.cn/en/cd/2003-11/14/content_281438.htm. (Accessed August 8, 2010).
39. "Ecotourism to Save Nature," November 14, 2003.
40. Jiang Shiwei, "Wanglang zai Min shan," August 6, 2010.
41. Local worker, interview by author, May 2002, Pingwu County, Sichuan, PRC.
42. "Duzi qu luyou 3—Zhongguo di 57 ge minzu" [Lone travel 3—China's 57th nationality] from Xiecheng http://www.ctrip.com/ found on

http://www.chinazijiayou.com/news/92/2007/129102756252.htm (Accessed August 8, 2010).

43. Wanglang reserve staff member, interview, July 1, 2013.
44. Binbin V. Li, Stuart L. Pimm, Sheng Lie, Lianjun Zhao, and Chunping Luo, "Free-Ranging Livestock Threaten the Long-Term Survival of Giant Pandas," *Biological Conservation* 216 (2017), 23–24.
45. Wang Hao, Ph.D., Lecturer, Peking University, School of Life Sciences, interview by author, August 9, 2016.
46. Wang Hao, interview, August 9, 2016.
47. Lü Zhi, personal correspondence, August 13, 2015.
48. Wang Hao, interview, August 9, 2016.
49. Wang Hao, interview, August 9, 2016.
50. Wang Hao, interview, August 9, 2016.
51. Henry Fountain, "In China Quake Analysis: Insight into Devastation," *New York Times*, Science, September 28, 2009. http://www.nytimes.com/2009/09/29/science/29obquake.html?rref=collection%2Ftimestopic%2FSichuan%20Earthquake&action=click&contentCollection=timestopics®ion=stream&module=stream_unit&version=search&contentPlacement=8&pgtype=collection
52. Wang Hao, interview, August 9. 2016.
53. Wanglang National Nature Reserve Web site, Home page, http://www.wanglang.com/, accessed January 14, 2013.
54. Wanglang Nature Reserve TEAM Network webpage, http://www.teamnetwork.org/en/field_stations/wanglang-nature-reserve (Accessed August 7, 2010) lists Wanglang as one of its few non-tropical scientific research sites in China.
55. John Seidensticker, John F. Eisenberg, and Ross Simons, "The Tangjiahe, Wanglang, and Fengtongzhai Giant Panda Reserves and Biological Conservation in the People's Republic of China," *Biological Conservation* 28, No. 3 (1984): 217–251; Alan H. Taylor, Qin Zisheng, Liu Jie, "Structure and Dynamics of Subalpine Forests in the Wang Lang Natural Reserve, Sichuan, China," *Vegetation* 124 (1996): 24–38; Zhan, Xiangjiang and others, "Molecular Censusing Doubles Giant Panda Population Estimate in Key Nature Reserve," *Current Biology* 16, no. 12 (2006): 451–452.
56. See Chapter 3: "The Winding Road to Wanglang: Creating a Panda Reserve" for more on the far reaches of nationalistic sentiment.

CHAPTER 8

1. "Panda Ambassador Mei Lan Ushers in Chinese New Year by Launching Earth Hour 2010 to the World," *WWF* (World Wide Fund for Nature) *Global*, February 11, 2010, http://wwf.panda.org/?188762 (Accessed January 21, 2012); World Wide Fund for Nature (WWF), Earth Hour Website: http://www.earthhour.org/ (Accessed on July 24, 2015).

2. World Wide Fund for Nature (WWF), Earth Hour Website: http://www.andymurray.com/news/andy-murray-serves-up-earth-hour-challenge-to-the-uk/; http://www.wwf.eu/?207918/actress-jessica-alba-announced-as-earth-hour-2013-global-ambassador (Accessed on July 24, 2015).
3. Huang Shiqiang, "Bu chuguo men de da xiongmao" [Stages in the process of giant pandas going abroad], unpublished essay, 2002, 2.
4. Huang Shiqiang, "Bu chuguo men de da xiongmao," 4.
5. "Pandas Extend Coast Visit," *The New York Times*, October 1, 1984, B11.
6. Schaller, *Last Panda*, 239–241.
7. Schaller, *Last Panda*, 235–249.
8. Even in 2009, after a great deal of success in captive breeding and there was a resurgent interest in reintroduction, such as described in F. Shen, Z. Zhang, W. He, B. Yue, A. Zhang, L. Zhang, R. Hou, C. Wang, and T. Watanabe, "Microsatellite Variability Reveals the Necessity for Genetic Input from Wild Giant Pandas (*Ailuropoda melanoleuca*) into the Captive Population," *Molecular Ecology* 18 (March 2009): 1061–1070. doi: 10.1111/j.1365-294X.2009.04086.x pointed to the challenges of sustainable captive breeding.
9. Xie and Gipps, *2001 International Studbook for Giant Panda*, 22–37.
10. CITES (Convention on Trade in Endangered Species of Wild Fauna and Flora) text, Washington, D.C., 1973, Amended at Bonn, 1979. http://www.cites.org/eng/disc/text.php#II; Appendices: http://www.cites.org/eng/app/appendices.php (Note: The giant panda is in Appendix I—under the greatest threat and strictest regulation. Accessed January 15, 2012).
11. CITES, http://www.cites.org/eng/disc/text.php#II; Appendices: http://www.cites.org/eng/app/appendices.php (Note: The giant panda is in Appendix I—under the greatest threat and strictest regulation. Accessed January 15, 2012).
12. Schaller, *Last Panda*, 237, 243.
13. Schaller, *Last Panda*, 243–248.
14. In Chinese the number eight is considered auspicious in part because it is a homonym with a Chinese word for fortune and prosperity.
15. Mu Xuequan, ed. "Six pandas arrive in Beijing to celebrate China's 60th anniversary," *China View* online, April 29, 2009. http://news.xinhuanet.com/english/2009-04/29/content_11283702.htm (Accessed August 2011). In China, both because of the patterns and cycles of the traditional Chinese lunar calendar and the more auspicious sound of the number six, the sixtieth birthday and sixtieth anniversary are more significant than the fiftieth and are thus celebrated in a similar fashion to fiftieth anniversaries of events in western countries.
16. Cui Lei, ed. "Zeng Xiang da xiongmao ni yu 4 yue 26 ri qicheng," *Renmin ribao* online, April 16, 2007. http://unn.people.com.cn/GB/14800/21806/5620841.html (Accessed, August 9, 2011).
17. "Two pandas selected to move to Macau," *Macau Daily Times* online, May 25, 2010. http://www.macaudailytimes.com.mo/ (Accessed May 25, 2010).

18. Xie and Gipps, *2001 International Studbook for Giant Panda,* 34–41.
19. John Watts, "1,3000 Years of Global Diplomacy Ends for China's Giant Pandas," *The Guardian International,* September 14, 2007, 27.
20. Schaller, *Last Panda,* 247–9.
21. Lauren Strapagiel, "Nishiyuu Journey by Cree Youth Ends as Harper Greets Pandas," *Huffington Post,* March 25, 2013, http://www.huffingtonpost.ca/2013/03/25/nishiyuu-journey-ends-ottawa-harper-pandas_n_2950643.html, (Accessed July 27, 2017); Emily Mertz, "Harper's Panda Meeting Sparks Criticism," *Global News,* March 25, 2013, http://globalnews.ca/news/427585/harpers-panda-meeting-sparks-criticism/ (Accessed August 21, 2017). Thank you to Joanna Hindle for calling my attention to this incident.
22. Enav, "Chinese Pandas Arrive in Taiwan," December 23, 2008.
23. Thomas Gold, "The Status Quo is Not Static: Mainland-Taiwan Relations," *Asian Survey* 27, No. 3 (March 1987), 302.
24. John Copper, "Taiwan in 1986: Back on Top Again," *Asian Survey* 27, No. 1 (January 1987), 88; "Report Finds Taiwan Leads in Economic Growth," *Journal of Commerce,* August 22, 1986.
25. "Peking Zoo Willing to Give a Pair of Giant Pandas to Zoo in Taiwan," *Xinhua News Agency,* April 10, 1987.
26. "Foreign News Briefs," *United Press International,* August 13, 1986.
27. "Taiwan Turns Down Mainland Panda Offer," *Japan Economic Newswire,* December 9, 1988.
28. Schaller, *Last Panda,* 244.
29. Free China Journal Editors, "Mainland–Taiwan Animal Exchange Being Studied," *The Free China Journal* 6, no. 29, April 24, 1989, 3.
30. Free China Journal Editors, "Taipei Panda Expertise Still Being Questioned," *The Free China Journal* 6, no. 31 (May 1, 1989), 3.
31. "Taiwan Rejects China's Panda Offer, *Reuters News,* July 23, 1990; "Taiwan says 'No' to China's Panda Offer," *Straits Times,* July 24, 1990.
32. Free China Journal Editors, "Panda Plan Withdrawn by City Zoo Officials," *The Free China Journal,* 7, no. 71 (September 17, 1990), 3.
33. "BBC Summary of World Broadcasts," *Xinhua News Agency,* March 11, 1995.
34. Fang Hsu and Victor Lai, "Taiwanese party delegates, ARATS officials discuss resuming cross-strait talks," *Taiwanese Central News Agency,* August, 31, 2000; Lilian Wu, "Pandas May Come to Taiwan," *Central News Agency* (Taiwan), June 6, 1997.
35. "Giant Panda Couple—Free Gifts for Taiwan," *Xinhua,* February 27, 2006, http://au.china-embassy.org/eng/xw/t237132.htm (Accessed on December 15, 2017).
36. "Names Unveiled for Panda Pair for Taiwan," *Xinhua,* January 28, 2006. news.xinhuanet.com/english/2006-01/28/content_4113317.htm.
37. Yang Meng-yu, "Taiwan wucheng minzhong huanying xiongmao daolai," [In Taiwan fifty percent of the populace welcome the arrival of pandas], *Lienhe bao,* January 16, 2006.

38. Richard Sobel, William-Arthur Haynes, and Yu Zheng, "The Polls—Trends: Taiwan Public Opinion Trends, 1992–2008: Exploring Attitudes on Cross-Strait Issues," *Public Opinion Quarterly* 74, No. 4 (Winter 2010): 782–813.
39. Ko Shu-ling, "Focus on Pandas, not Missiles: Chen," *Taipei Times*, March 31, 2006.
40. Liao Hongxiang, "Da maoxiong de xianjing" [Giant panda trap], *New Taiwan*, May 5, 2005, http://www.newtaiwan.com.tw/bulletinview.jsp?bulletinid=21901 (Accessed on July 24, 2010).
41. Associated Press, "Taiwan rejects China's Offer of Pandas," *USA Today*, March 31, 2006.
42. "Taipei Zoo Challenges COA's Panda Rejections," *China Post*, June 25, 2007; "Zoo Sticks to its Guns over Accepting China's Pandas," *China Post*, March 28, 2008.
43. Chen Shui-pien, "Born Free," A-Pien Zongtong Dianzi Youbao [President Pien's Blog], http://www.president.gov.tw/1_president/subject-05.html, March 23, 2006 (Accessed February 2008).
44. Council of Agriculture, Executive Yuan, "Da maoxiong anzhang weihui xikai di san ci pancha hui" [Third meeting on the application for giant pandas], 1995 nian 3 yue 31 ri, March 31, 2006 (Special note: Taiwan's system for dates is based on the notion that the year 1911 is the first year of the Republic. Consequently it is officially considered to be the first year, which effectively subtracts eleven years from the date. Thus, a date that reads 1995 is equivalent to 2006 according to standard western calendars).
45. Council of Agriculture, Executive Yuan, "Da maoxiong anzhang," March 31, 2006.
46. "Taiwan Rejects China's Giant Pandas," *IOL*, March 31, 2006; "Government Rejects China's Offer for Gift Pandas," *China Post*, April, 1, 2006.
47. Yang Li, ed., "Taiwan Rejects Pandas," *China Daily*, April 1, 2006, www.chinaview.cn 2006-04-01 09:37:05 (accessed July 21, 2010).
48. "Are Pandas Spies?" *People's Daily*, April 5, 2006, http://english.people.com.cn/ (accessed July 21, 2010). Please note, the PRC newspaper, *People's Daily* uses the pinyin system to spell Chen Shui-pien.
49. Yang Li, "Taiwan Rejects Pandas," *China Daily*, April 1, 2006.
50. "Chen Justifies Refusing Gift Pandas," *The China Post*, April 7, 2006. http://www.chinapost.comtw/print/79963.htm (Accessed on July 21, 2010).
51. Yang Meng-yu, "Xiongmao mei lai; kao ya lai," *BBC Chinese.com*, July 6, 2007, http://newsvote.bbc.co.uk/mpapps/pagetools/pring/news.bbc.co.uk/ch (Accessed on July 22, 2010.)
52. Rigger, "Taiwan's Presidential and Legislative Elections," 689.
53. "China Renews Panda Offer," *Taipei Times*, February 1, 2008, 2.
54. Sophie Yu, "Taiwan to Accept Pandas," *The Times* (London), March 24, 2008.
55. "Chinese Mainland Hopes to Send Panda Pair to Taiwan Soon," *Xinhua*, January 31, 2008.

56. Rigger, "Taiwan's Presidential and Legislative Elections," 692; Raju Gopalakrishnan and Jonathan Standing, "Taiwan Says Yes to Ma Re-Election and His 'Three No's,'" *Reuters*, January 15, 2012.
57. Mo Yan-chih, "Panda Cub Officially Named Yuan Zai, Gets Citizen's Card," *Taipei Times*, October 27, 2013. http://www.taipeitimes.com/News/taiwan/archives/2013/10/27/2003575496 (Accessed on July 30, 2015).
58. Scarlett Chai, Ku Chuan and Jay Chen, "Chen Deming Visits Panda Cub as He Wraps Up Taiwan Trip," *Focus Taiwan News Channel*, February 28, 2014. http://focustaiwan.tw/search/201402280007.aspx?q=visit%20panda%20cub (Accessed July 30, 2015).

CONCLUSION

1. Nathan Yaussy, creator of EUT, Endangered Ugly Things, website, posted purpose of website. http://endangered-ugly.blogspot.com/ (Accessed December, 18, 2010). This site has moved to a new location: http://endangereduglythings.tumblr.com/, which has archives since 2014.
2. Ben Blanchard, editing by Bill Tarrant, "Panda Attacks Man Who Wanted a Cuddle," *Reuters*, November 24, 2008. http://www.reuters.com/article/us-china-panda-idUSTRE4AN5NF20081124; "Panda Attacks Man in Chinese Zoo," BBC News, November 22, 2008, http://news.bbc.co.uk/2/hi/7743748.stm (Accessed August 21, 2017).
3. Carla Hall, "National Zoo Panda Cam Shuts Down! Bad Idea!" *Los Angeles Times*, October 1, 2013. http://articles.latimes.com/2013/oct/01/news/la-ol-national-zoo-panda-cam-shuts-down-20131001; Gregory Wallace, "'Panda Cam' Goes Dark in Shutdown," *CNN, Money*, October 1, 2013. http://money.cnn.com/2013/09/30/news/panda-cam-national-zoo/index.html (Accessed August 21, 2017).
4. Blogger on San Diego Zoo site: November 8, 2012, http://blogs.sandiegozoo.org/2012/11/06/exam-12-confident-and-curious/#comments (Accessed November 19, 2012).
5. Blogger on San Diego Zoo site: December 1, 2012, http://blogs.sandiegozoo.org/2012/11/29/panda-cub-exam-15/ (Accessed December 3, 2012).
6. David Gray, "A Necessary Evil—the Kangaroo Cull," *Photographer's Blog* on *Reuters* http://blogs.reuters.com/photographers-blog/2013/04/03 /a-necessary-evil-the-kangaroo-cull/, April 3, 2013 (Accessed August 21, 2017); "Hunting Kangaroos," *Aussiehunter*, https://aussiehunter.org/hunting/where-to-shoot/hunting-kangaroos/ (Accessed August 21, 2017).

Bibliography

Allen, Sara. *The Shape of the Turtle: Myth, Art, and Cosmos in Early China.* New York: State University of New York Press, 1991.

Andrews, Julia F. *Painters and Politics in the People's Republic of China, 1949–1979.* Berkeley: University of California Press, 1994.

Andrews, Julia F., and Kuiyi Shen. *A Century in Crisis: Modernity and Tradition in the Art of Twentieth-Century China.* New York: Guggenheim Museum, 1998.

"Are Pandas Spies?" *People's Daily,* April 5, 2006, http://english.people.com.cn/. Accessed July 21, 2010.

Associated Press. "Taiwan Rejects China's Offer of Pandas." *USA Today,* March 31, 2006.

"Baima Zangzu wei Dizu shuo zhiyi." (The Baima Tibetans and the Di people of Chinese historical records: Challenging the link). *Bulletin of Chinese Linguistics,* 3, no. 1 (2008): 167–180.

"BBC Summary of World Broadcasts." *Xinhua News Agency,* March 11, 1995.

Becker, Jasper. *Hungry Ghosts: Mao's Secret Famine.* New York: Henry Holt and Company, 1996.

"Beijing dongwu yuan da xiongmao chanzi" (A giant panda is born at the Beijing Zoo). *Renmin ribao (People's Daily),* 28 November 1963, 2.

"Beijing shi gewei hui daizhuren Wu De huijian husong sheniu de meiguo keren" (Beijing municipal revolutionary committee representative Wu De meets and escorts muskoxen guests from America). *Renmin ribao (People's Daily),* April 13, 1972.

"Beijing shi gewei hui zeng songgei meiguo ren min de yi dui xiongmao zai huashengdun guojia dongwuyuan juxing jiaojie yishi" (A ceremony was given in Washington D.C. at the National Zoo for the pair of pandas that the Beijing municipal revolutionary committee presented the American people). *Renmin ribao (People's Daily),* April 22, 1972, 6.

Bhanoo, Sindya N. "Ancient Bear May Be Ancestor of Giant Panda." *New York Times,* November 19, 2012.

Blanchard, Ben, editing by Bill Tarrant, "Panda attacks man who wanted a cuddle," *Reuters*, November 24, 2008. http://www.reuters.com/article/us-china-panda-idUSTRE4AN5NF20081124. Accessed October 16, 2017.

Carey, Geoff, ed. *A Biodiversity Review of China*. Hong Kong: World Wide Fund for Nature (WWF) International, 1996.

Catten, Chris. *Pandas*. London: Christopher Helm, 1990.

Chai, Scarlett, Ku Chuan, and Jay Chen. "Chen Deming Visits Panda Cub as He Wraps Up Taiwan Trip." *Focus Taiwan News Channel*, February 28, 2014. http://focustaiwan.tw/search/201402280007.aspx?q=visit%20panda%20cub. Accessed July 30, 2015.

Chan, Anita, Richard Madsen, and Jonathan Unger. *Chen Village: The Recent History of a Peasant Community in Mao's China*. Berkeley: University of California Press, 1984.

Chen, Guidi, and Wu Chuntao. *Will the Boat Sink the Water: The Life of China's Peasants*. New York, Public Affairs, 2007.

"Chen Justifies Refusing Gift Pandas." *The China Post*, April 7, 2006. http://www.chinapost.comtw/print/79963.htm. Accessed on July 21, 2010.

Chen, Liang "Ecotourism to Save Nature." *China Daily*, November 14, 2003. http://www.chinadaily.com.cn/en/cd/2003-11/14/content_281438.htm. Accessed August 8, 2010.

Chen, Shui-pien. "Born Free." A-Pien Zongtong Dianzi Youbao (President Pien's Blog), http://www.president.gov.tw/1_president/subject-05.html, March 23, 2006. Accessed February 2008.

Chen, Yuanfei. "Da xiongmao 1983 dashi ji" (Major Giant Panda Events in 1983). *Da ziran tansuo* (Examining Nature) 2, no. 8 (1984): 185.

Cheung, Siu-woo. "Representation and Negotiation of Ge Identities in Southeast Guizhou." In *Negotiating Ethnicities in China and Taiwan*, edited by Melissa J. Brown. Berkeley: Institute for East Asian Studies, 1996, 240–273.

"Chicago Hails China Troupe." *New York Times*, December, 23 1972, 15.

China Pictorial (Renmin huabao人民画报). 1966–1969, 1973.

"China Plans to Launch 4th National Survey on Giant Pandas in 2011." *People's Daily Online*, April 1, 2011, http://english.peopledaily.com.cn/90001/90782/6938040.html. Accessed July 7, 2010.

"China Presents Japan with 2 Giant Pandas." *New York Times*, September 29, 1972.

"China Renews Panda Offer." *Taipei Times*, February 1, 2008, 2.

"Chinese Mainland Hopes to Send Panda Pair to Taiwan Soon." *Xinhua*, January 31, 2008.

Chirkova, Katia. "Between Tibetan and Chinese: Identity and Language in Chinese South-West." *Journal of South Asian Studies* 30, no. 3(2007): 405–417.

"Chuantong gongyi kai xinhua" (Updating Traditional Art and Craft). *Renmin ribao* (*People's Daily*), June 5, 1973, 4.

Chung, Lawrence. "Taiwan Says No Thanks to Beijing's Panda Offer." *South China Morning Post*, April 1, 2006.

Cihai. Shanghai: Shanghai si chu chuban she, 1989.

Ciochon, Russell L. Professor and Chair, Department of Anthropology, University of Iowa. Personal correspondence, December 10 and November 21, 2008.

CITES (Convention on Trade in Endangered Species of Wild Fauna and Flora) text. Washington, D.C., 1973. Amended at Bonn, 1979. http://www.cites.org/eng/disc/text.php#II. Accessed January 15, 2012.

CITES (Convention on Trade in Endangered Species), Appendices. http://www.cites.org/eng/app/appendices.shtml. Accessed January 15, 2012.

Clunas, Craig. *Art in China*. Oxford: Oxford University Press, 1997.

Copper, John. "Taiwan in 1986: Back on Top Again." *Asian Survey* 27, no. 1 (January 1987): 81–91.

Council of Agriculture, Executive Yuan. "Da maoxiong anzhang weihui xikai di san ci pancha hui" [Third meeting on the application for giant pandas], 1995 nian 3 yue 31 ri, March 31, 2006.

Cui, Lei, ed. "Zeng Xiang da xiongmao ni yu 4 yue 26 ri qicheng." *Renmin ribao* online, April 16, 2007. http://unn.people.com.cn/GB/14800/21806/5620841.html. Accessed August 9, 2011.

Da xiongmao ziran siwang lianhe diaocha dui (Giant panda natural death united survey team). Survey Report, July, 20, 1976.

David, Armand. *Journal de mon troisieme d'exploration dans l'empire de Chinois par l'Abbe Armand David*. Paris: Hachette, 1875.

David, Armand. "Journal d'u voyage dans le center de la Chine et dans le Thibet Oriental." *Bulletin Nouvelle Archives Muséum d'Histoire Naturelle de Paris* 10: 3–82.

David, Armand. *Abbé David's Diary: Being an Account of the French Naturalists's Journeys and Observations in China in the Years 1866-1869*. Translated by Helen M. Fox. Cambridge, Mass.: Harvard University Press, 1949.

Davis, D. Dwight. *The Giant Panda: A Morphological Study of Evolutionary Mechanisms*. Chicago: Chicago Natural Museum, 1964.

Davis, Joseph A. "A China Doll and Pal Come to Town." *New York Times*, April 23, 1972, E5.

Dementiev, G. P. "Sulian de ziran baohu" (Nature protection in the Soviet Union). *Dongwuxue zazhi* 1.4 (1957): 210–211.

Dong, Zhiyong. "Qiangjiu 'guobao' da xiongmao" (Saving the 'national treasure' the giant panda). *Yesheng dongwu* (Chinese Wildlife) 3 (May 1984): 1–3.

Dong, Zhiyong, ed. *Zhongguo linye nianjian* (China Forestry Yearbook). Beijing: Zhongguo linye chuban she, 1987.

"Dongjing juxing jieshou Zhongguo renmin zengsong yi dui da xiongmao yishi Erjie tang jin guanfang zhangguan, Qiaoben deng mei sanlang ganshi zhang, yi ji ge jie renshi canjia" (Tokyo held a reception for the pair of pandas given by the Chinese people, Commanding Officer, Head Secretary, and other public figures from other divisions all participated). *Renmin ribao* (*People's Daily*), November 5, 1972, 4.

Dunlap, Thomas. *Saving America's Wildlife*. Princeton: Princeton University Press, 1988.

"Duzi qu luyou 3—Zhongguo di 57 ge minzu" (Lone travel 3—China's 57th nationality) from Xiecheng http://www.ctrip.com/ found on http://www.chinazijiayou.com/news/92/2007/129102756252.htm. Accessed August 8, 2010.

Edmonds, Richard L. *Patterns of China's Lost Harmony: A Survey of the Country's Environmental Degradation and Protection*. London: Routledge, 1994.

Enav, Peter. "Chinese Pandas Arrive in Taiwan in Charm Offensive." *Associated Press*, December 23, 2008.

Fan, Fa-ti. "Redrawing the Map: Science in Twentieth-Century China." *Isis* 98 (2007): 524–538.

Farnsworth, Lee W. "Japan 1972: New Faces and New Friends." *Asian Survey* 13, no. 1 (January, 1973): 122.

Faulkenheim, Victor. "Continuing Central Predominance." *Problems of Communism* 21, no. 4 (1972): 75–83.

"Fazhan qing gongye yuanliao jidi Nei Menggu qing gongye bumen jianli jidi zhengqu yuanliao zhubu zi gei Rui Anbai hao ruping chang tongguo zhiyuan xumuye fazhan guang bi ru yuan" (The basic materials for the development of light industry. The departments involved in Inner Mongolia's light industry established basic materials to pursue the development of milk product production to support the development of husbandry and the expansion of milk resources). *Renmin ribao* (*People's Daily*), February 17, 1960, 2.

Feng, Wenhe, and Li Guanghan. *Zhengjiu da xiongmao* (Saving the Giant Panda) Chengdu: Sichuan Kexue jishu chubanshe, 2000.

Fernsebner, Susan. "Material Modernities: China's Participation in World's Fairs and Expositions, 1876–1955." Ph.D., diss., University of California, San Diego, 2002.

"Foreign News Briefs." *United Press International*, August 13, 1986.

Fountain, Henry. "In China Quake Analysis: Insight into Devastation." *New York Times*, Science, September 28, 2009. URL: http://www.nytimes.com/2009/09/29/science/29obquake.html?rref=collection%2Ftimestopic%2FSichuan%20Earthquake&action=click&contentCollection=timestopics®ion=stream&module=stream_unit&version=search&contentPlacement=8&pgtype=collection. Accessed October 16, 2017.

Free China Journal Editors. "Mainland–Taiwan Animal Exchange Being Studied." *The Free China Journal* 6, no. 29 (April 24, 1989), 3.

Free China Journal Editors. "Taipei Panda Expertise Still Being Questioned." *The Free China Journal* 6, no. 31 (May 1, 1989), 3.

Free China Journal Editors. "Panda Plan Withdrawn by City Zoo Officials." *The Free China Journal*, V. 7, no. 71, September 17, 1990, 3.

Fu, Yingquan, ed. *Sichuan zhuan* (Sichuan volume). *Zhongguo ziran ziyuan cong shu* (Chinese natural resources series) no. 33. Beijing: Zhongguo huanjing kexue chuban she, 1995.

Gansu sheng zhengui dongwu ziyuan diaocha dui, "Gansu de da xiongmao," *Lanzhou daxue xuebao* 3 (1977): 88–99.

Garshelis, David, Wang Hao, Wang Dajun, Zhu Xiaojian, Li Sheng, William J. McShea., "Do Revised Giant Panda Population Estimates Aid in their Conservation?" *Ursus* 19, no. 2 (2008): 168–176.

Geological Society of America GSA. "Geologic Time scale." http://www.geosociety.org/science/timescale/. Accessed November 2010.

"Giant panda couple—free gifts for Taiwan." *Xinhua*, February 27, 2006, http://au.china-embassy.org/eng/xw/t237132.htm.

"'Giant Panda Originates in Europe' Incredible [*sic*]." *Beijing Review*, 41, no. 22 (1–7 June 1998): 28.

Gold, Thomas. "The Status Quo Is Not Static: Mainland-Taiwan Relations." *Asian Survey* 27, No. 3 (March 1987): 300–315.

Goodman, David. *Centre and Province in the PRC: Sichuan and Guizhou, 1955–1965*. Cambridge, Cambridge University Press, 1986.

Gopalakrishnan, Raju, and Jonathan Standing. "Taiwan Says Yes to Ma Re-Election and His 'Three No's.'" *Reuters*, January 15, 2012.

Gould, Stephen J. "The Panda's Thumb." In *The Panda's Thumb: More Reflections in Natural History*. New York: W. W. Norton and Company, 1980.

Gould, Stephen J. "How Does the Panda Fit?" In *An Urchin in the Storm*, New York: W. W. Norton and Company, 1987.

"Government Rejects China's Offer for Gift Pandas." *China Post*. April 1, 2006.

Gross, Michael. Cover Art, *National Lampoon* (July 1972).

Grauwels, Stephan. "Taiwan Rejects China's Offer of Pandas." Associated Press Online, March 31, 2006 in http://www.lexisnexis.com.ezproxy1.lib.ou.edu/us/lnacademic/results/docview/docview.do?docLinkInd=true&risb=21_T5500773876&format=GNBFI&sort=BOOLEAN&startDocNo=1&resultsUrlKey=29_T5500773879&cisb=22_T5500773878&treeMax=true&treeWidth=0&csi=147876&docNo=1. Accessed January 9, 2009.

Gray, David. "A Necessary Evil—the Kangaroo Cull." *Photographer's Blog* on *Reuters*, April 3, 2013, http://blogs.reuters.com/photographers-blog/2013/04/03 /a-necessary-evil-the-kangaroo-cull/, Accessed August 21, 2017.

Guo, An. "Wo guo zui da de dongwu yuan—Beijing dongwu yuan" (Our nation's largest zoo—the Beijing Zoo). *Renmin ribao* (*People's Daily*), May 6, 1956, 3;

Guangming ribao. May 5, 1999.

Gujin tushu jicheng. (Compilation of books and illustrations, past and present.) Taipei: Tingwen shu chu, 1977 [1725].

Guojia huanjing baohu ju ziran baohu si (National environmental protection agency). *Ziran baohu qu youxiao guanli lunwen ji* (Effective management of nature reserves, a collection of essays). Beijing: Zhongguo huanjing kexue chubanshe, 1992.

Hall, Carla. "National Zoo Panda Cam Shuts Down! Bad idea!" *Los Angeles Times*, October 1, 2013. http://articles.latimes.com/2013/oct/01/news/la-ol-national-zoo-panda-cam-shuts-down-20131001. Accessed August 21, 2017.

Han, Zhengfu. "Han Zhengfu tongzhi zai quan sheng zhengui dongwu ziyuan, diaocha zuotan huiyi shang de jianghua," (Comrade Han Zhengfu's speech at the Province-wide meeting on precious animal resources and surveys). Delivered at Sichuan's Province-wide meeting on precious animal resources and surveys (November 17, 1973).

Hao, Siyong. "Xiongmao wu zhou sui" ('Panda' is five years old). *Renmin ribao (People's Daily)*, September 30, 1961, 5.

Harkness, Ruth. *The Lady and the Panda.* London: Nicholson and Watson, 1938.

Harper, Donald. "The Cultural History of the Giant Panda (Ailuropoda melanoleuca) in Early China." *Early China* 35–36 (2012–2013): 185–224.

Harrell, Stevan. "The Nationalities Question and the Prmi Problem." In *Negotiating Ethnicities in China and Taiwan,* edited by Melissa J. Brown, 274–296. Berkeley: Institute for East Asian Studies, 1996.

Harrell, Stevan. "The History of the History of the Yi." In *Cultural Encounters on China's Ethnic Frontiers,* edited by Stevan Harrell. Seattle: University of Washington Press, 1995.

Harrell, Stevan. "Introduction." In *Cultural Encounters on China's Ethnic Frontiers.* Seattle: University of Washington Press, 1996.

Hashimoto, Tetsuo, Eiko Otaka, Jun Adachi, Keiko Mizuta, Masami Hasegawa. "The Giant Panda Is Closer to a Bear, Judged by α- and β-Hemoglobin Sequences." *Journal of Modern Evolution* 36 (1993): 282–289.

Hatena Diary blog, http://d.hatena.ne.jp/syulan/20070523/p1. Accessed March 4, 2008.

Hathaway, Michael. "The Emergence of Indigeneity: Public Intellectuals and the Indigenous Space in Southwest China." *Cultural Anthropology* 25, no. 2 (March 5, 2010): 301–333.

Hays, Samuel P. *Conservation and the Gospel of Efficiency: The Progressive Conservation Movement, 1890–1920.* New York: Atheneum, 1979 [1959].

He, Daihua, ed. *Ke'ai de jia xiang Pingwu* (Our endearing hometown of Pingwu). Mianyang: Mianyang shi weicheng caiying chang ying shua, 1998.

He, Guangming, Xiaodong Chen, Wei Liu, Scott Bearer, and Shiqiang Zhou. "Distribution of Economic Benefits from Ecotourism: A Case Study of Wolong Nature Reserve for Giant Pandas in China." *Environmental Management* 42 (2008): 1017–1025.

Hempson-Jones, Justin S. "The Evolution of China's Engagement with International Governmental Organizations: Toward a Liberal Foreign Policy." *Asian Survey* 45, no. 5 (2005): 702–721.

Higgins, Andrew. "Taiwan Unlikely to Move to Reunify with China, Despite Ma Ying-jeou's Reelection." The *Washington Post with Foreign Policy,* January 15, 2012. http://www.washingtonpost.com/world/asia_pacific/taiwan-wants-a-separate-peace-with-china/2012/01/15/gIQA3ufF1P_story.html. Accessed January 20, 2012.

Hineline, Mark. Lecture. "The Built Environment." San Diego, University of California, 2000.

Ho, Dahpon. "To Protect and Preserve: Resisting the 'Destroy the Four Olds' Campaign, 1966–1967." Presented at the UCSD-Stanford Cultural Revolution Conference June, 8–9 2003, University of California, San Diego.

Hsing, You-tien. "Brokering Power and Property in China's Townships." *The Pacific Review* 19, no. 1 (2006): 103–124.

Hsu, Fang, and Victor Lai. "Taiwanese Party Delegates, ARATS Officials Discuss Resuming Cross-Strait Talks." *Taiwanese Central News Agency*, August, 31, 2000.

Hsü, Immanuel C. Y. *China Without Mao: The Search for a New Order*, 2nd ed. New York: Oxford University Press, 1990.

Hu, Jinchu. "Canjia Zhongguo dongwuxue hui chengli wushi zhou nian nian hui de dongwuxue gongzuozhe huyu: caiqu jinji jieshi baohu da xiongmao" (Participating in the fiftieth annual meeting of China's national zoological meeting, a zoological worker's appeal: take urgent measures to protect the giant panda). *Yesheng dongwu* (Chinese wildlife) 5 (September 1984): 1–3.

Hu, Jinchu. *Da xiongmao de yanjiu* (Research on the giant panda). Shanghai: Shanghai keji yiaoyu chuban she, 2001.

Hu, Jinchu, George B. Schaller, Pan Wenshi, and Zhu Jing. *Wolong de da xiongmao* (The giant pandas of Wolong). Chengdu: Sichuan kexue jishu chuban she, 1985.

Hu, Tieqing, former head of Sichuan Provincial Wild Animal Protection Department. Interview by author, Chengdu, Sichuan Province, PRC, January 14, 2002.

"Hu yin shanlu gong youyi—ji di shi ci ri zhong youhao xingnian xialing ying" (Home in the hidden foothills praising friendship—records of the tenth Japanese-Chinese friendship summer camp for young people). *Renmin ribao* (*People's Daily*), August 8, 1973, 5.

Huang, Ellen. "Jingdezhen Porcelain: Producing china and China." Presented at University of California, Berkeley, December 3, 2009.

"Huang, Hua. Wei wo ping pang qiu daibiao tuan juxing shaodai hui dui Meiguoren he pingxie deng gei yu daibiaotuan reqing huanying he youyu gao jiedai biaoshi ganxie" (Huang Hua acts as our ping pong team representative and holds a reception to express gratitude to the American people and the ping pong association for giving our team a warm welcome and friendship). *Renmin ribao* (*People's Daily*), April 23, 1972, 6.

Huang, Shiqiang. "Bu chuguo men de da xiongmao" (Stages in the process of giant pandas going abroad). Unpublished essay, 2002.

"Hubei shou gongyi lao yiren qizuo yipi xin zuopin" (Old artists create new art in Hubei handicrafts). *Renmin ribao* (*People's Daily*), August 27, 1972, 3.

"Hunting Kangaroos." *Aussiehunter*. https://aussiehunter.org/hunting/where-to-shoot/hunting-kangaroos/ Accessed August 21, 2017.

Integrated Conservation Development Program (ICDP). "Pingwu xian da xiongmao qixidi zonghe baohu yu fazhan xiangmu, Baima zang zu xiang shehui jingji diaocha baogao" (Pingwu county giant panda habitat Integration of Conservation and Development Program, Baima Tibetan community social and economic survey report), June 1998.

IUCN (International Union for the Conservation of Nature) Red List of Threatened Species, http://www.iucnredlist.org/apps/redlist/details/712/0. Accessed July 7, 2010.

Jiang, Shiwei. "Wanglang: Zai minshan shenchu jingjing zhanfang (Wanglang: the quiet blossom deep in the heart of the Min Mountains)." *Zhongguo luse shibao China's Green Times*, August 6, 2010. http://www.forestery.gove.cn/portal/main's'72/content-434604.html. Accessed August 7, 2010.

Jiang, Tingan. "Zai da xiongmao de guxiang" (In panda country). *Bowu* (Natural history) 3 (November 1980): 15–16.

"Jiji fazhan chuantong yigong meishu pin shenchan" (Enthusiastic development of traditional craft and artwork production). *Renmin ribao* (*People's Daily*) October 26, 1972, 2.

Jin, Changzhu, Russell L. Ciochon, Wei Dong, Robert M. Hunt, Jr., Marc Jaeger, and Qizhi Zhu. "The First Skull of the Earliest Giant Panda." Proceedings of the National Academy of Sciences 104, no. 26 (June 26, 2007): 10932–10937.

Jin, Jianming, ed. *Ziran baohu gailun* (An introduction to nature protection). Beijing: Zhongguo huanjing kexue chuban she, 1991.

Jin, Jianming, Wang Liqiang, and Xue Dayuan, eds. *Ziran baohu gailun* (An introduction to nature protection). Beijing: Zhongguo huanjing kexue chuban she, 1991.

Jin, Jieliu. "Guanyu 'ziran baohu'" (Concerning 'nature protection'). *Dongwuxue zazhi* (Journal of zoology) 1, no. 3(1957): 129–131.

Johnson, Chalmers. "The Patterns of Japanese Relations with China, 1952–1982." *Pacific Affairs* 59, no. 3 (Autumn, 1986): 402–428.

Johnson, Kenneth G., George B. Schaller, and Hu Jinchu. "Response of Giant Pandas to a Bamboo Die-off." *National Geographic Research* 4 (1988): 161–177.

Johnson, Matthew David. "A Politics of Form: Cultural Conflict and China's Film Industry After the Great Leap Forward, 1959–1964." Presented at the UCSD-Stanford Cultural Revolution Conference 8–9 June 2003, University of California, San Diego.

Joravsky, David. *The Lysenko Affair*. Chicago: Chicago University Press, 1986 [1970].

"Joint Communiqué on the Establishment of Diplomatic Relations with the People's Republic of China." January 1, 1979, Embassy of the People's Republic of China in the United States of America. http://www.china-embassy.org/eng/zmgx/zywj/t36256.htm. Accessed July 31, 2010.

Kachur, Torah. "Is China the World's New Scientific Superpower?" *CBC News*, Sept 1, 2016. http://www.cbc.ca/news/technology/china-science-superpower-1.3743556. Accessed January15, 2017.

Kane, Penny. *Famine in China 1959–1961: Demographic and Social Implications*. Hong Kong: The Macmillan Press, 1988.

KarmaQuest, Adventure, Travel and Outdoor Learning that Support Conservation, web site http://www.karmaquests.com/index.htm. Accessed January 30, 2012.

Kim, Samuel S. "The People's Republic of China in the United Nations: A Preliminary Analysis." *World Politics* 26, no. 3 (April 1974), 301.

Ko, Shu-ling. "Focus on Pandas, Not Missiles: Chen." *Taipei Times*, March 31, 2006.

Kojima, Noriyuki et al., eds. *Nihon Shoki* 3. Tōkyō: Shōgakkan, 1994–1998.

Laidler, Keith, and Liz Laidler. *Pandas*. London: BBC Books, 1992.

Laluosikaya, N. [Chinese transliteration of Russian name]. "Ziran jieli meiyou mimi" (Nature has no secrets). *Zhishijiushi liliang* (Knowledge is power) (1958): 17.

Lamb, Christina. "Peevish China recalls panda Tai Shan due to Obama's meeting with the Dalai Lama." *The Sunday Times* (UK), February 14, 2010.

Lankester, E. R. "On the Affinities of *Æluropus malanoleucus*." *Linn. Society London* 2, no. 8 (1901): 163–171.

Lardy, Nicholas R. "Economic Recovery, 1963–1965." In *The Cambridge History of China, Volume 14, The People's Republic, Part I: The Emergence of Revolutionary China, 1949–1965*, ed. Denis Twitchett and John K. Fairbank. Cambridge: Cambridge University Press, 1987.

Lardy, Nicholas R. *Economic Growth and Distribution in China*. Cambridge: Cambridge University Press, 1978.

Laufer, Berthold. *Jade. A Study in Chinese Archeology and Religion*. New York: Dover Publishing, 1974.

Li, Binbin V., Stuart L. Pimm, Sheng Lie, Lianjun Zhao, and Chunping Luo. "Free-Ranging Livestock Threaten the Long-Term Survival of Giant Pandas." *Biological Conservation* 216 (2017): 18–25.

Li, Shaoming. "Zhongguo Qiangzu yu Baima Zangzu wenhua bijiao yanjiu" (Comparative research on China's Qiang and Baima Tibetan ethnic groups). In *Baima Zangzu yanjiu wenji*, edited by Zeng Weiyi (Chengdu: Sichuan sheng minzu yanjiu suo chuban, 2002), 183–193.

Li, Shengzhi, ICDP representative. Interview by author, May 25, 2002, Pingwu City, Pingwu County, Sichuan, PRC.

Li, Shizhen. *Ben cao gang mu* (Materia medica). Shanghai: Shanghai guji chuban she, 1991 [1596].

Li, Wenhua. *Zhongguo de ziran baohu qu* (China's nature reserves). Beijing: Shanwu ying shu guan, 1995.

Li, Xingxing, Professor of Anthropology, Sichuan Nationalities Institute. Interview by author, November 18, 2000, Orange County, California.

Liang you (Young companion) 153, no. 31 (April 1940): 42.

Liao, Hongxiang. "Da maoxiong de xianjing" [Giant panda trap]. *New Taiwan*, May 5, 2005, http://www.newtaiwan.com.tw/bulletinview.jsp?bulletinid=21901. Accessed on July 24, 2010.

Lieberthal, Kenneth. *Governing China: From Revolution Through Reform*. New York: W. W. Norton & Company, 1995.

Lieberthal, Kenneth, and Michel Oksenberg. *Policy Making in China: Leaders, Structures, and Processes*. Princeton: Princeton University Press, 1988.

Lin, Zhi. "China Marks 140th Anniversary of the Giant Panda's 'Discovery'." *China View*, August 15, 2009. http://news.xinhuanet.com/english/2009-08/15/content_11888057.htm. Accessed June 23, 2011.

Lindburg, Donald, and Karen Baragona, eds. *Giant Pandas: Biology and Conservation*. Berkeley: University of California Press, 2004.

Liu, Qiaofei. "Qianshi dongwu qu xide diaocha yanjiu" (Preliminary animal reserve system survey and research). *Dongwuxue zazhi* (Journal of Zoology) 3, no. 6 (1959): 287.

Liu, Xiangji, and Huang Zhenggen. "Baohu da xiongmao deng zhenxi dongwu yesheng dongwu baohu xiehui kaizhan mujuan huodong" (Protecting the giant panda and other precious and rare animals, the Wild Animal Protection Association develops fundraising activity). *Renmin ribao* (*People's Daily*), January 9, 1984, 3.

Liu, Yuanyun. "Chengdu lujun xuexiao wei qiangjiu da xiongmao yongyue juankuan juanliang" (The Chengdu army academy enthusiastically contributed money and grain to save the giant panda). *Sichuan dongwu* (Sichuan animals) 3, no. 3 (August 1984): 46.

Lü, Zhi, Lin Lin, Han Wei, Gan Tingyu, Li Shengzhi, and Xu Wei. "Pingwu da xiongmao qixidi baohu yu fazhan xiangmu, shehui jingji dioacha zong baogao" (Pingwu giant panda habitat protection and Integrated Community Development Program, community economic survey report), 1997.

Luo, Chunping. Sichuan Wanglang Guojia ji ziran baohu qu guanli ju (Wanglang National Nature Reserve management office). *Wanglang xinxi* (Wanglang news) no. 42 (June 2010): 2–3.

MacFarquhar, Roderick. *The Origins of the Cultural Revolution, vol. 1: Contradictions Among the People*. New York: Columbia University Press, 1974.

MacMillan, Margaret. *Nixon and Mao: The Week that Changed the World*. New York: Random House Trade Paperbacks, 2008.

Magnier, Mark. "Attack of the Pandas." *Los Angeles Times*, March 21, 2006.

Mao, Zedong. "Talks at the Yenan Forum on Literature and Art." *Mao Tse-tung on Literature and Art*. Beijing: Foreign Languages Press, 1977 [1960].

Markham-Smith, Ian. "US Zoo to Be Test Case for Panda Plan." *South China Morning Post* (Hong Kong), January 22, 1995.

Matthew, W. D., and Walter Granger, "Fossil Mammals from the Pliocene of Szechuan, China." *Bulletin of the American Museum of Natural History* 48, no. 17 (December 10, 1923): 563–598.

Matthews, Jay. "The Strange Tale of American Attempts to Leap the Wall of China." *New York Times*, April 18, 1971, 21.

Mendelsohn, Everett, Merritt Roe Smith, and Peter Weingart. *Science Technology and the Military*. Dordrecht: Kluwer Academic Publishers, 1988.

Menzies, Nicholas. *Forest and Land Management in Imperial China*. New York: St. Martin's Press, 1994.

Mo, Yan-chih. "Panda Cub Officially Named Yuan Zai, Gets Citizen's Card." *Taipei Times*, October 27, 2013. http://www.taipeitimes.com/News/taiwan/archives/2013/10/27/2003575496. Accessed on July 30, 2015.

Morris, Ramona, and Desmond Morris. *Men and Pandas*. New York: McGraw-Hill Book Company, 1966.

Morris, Ramona, and Desmond Morris. *The Giant Panda*. London: Kogan Page, 1981.

Mu, Xuequan, ed. "Six pandas arrive in Beijing to celebrate China's 60th anniversary." *China View* online, April 29, 2009. http://news.xinhuanet.com/english/2009-04/29/content_11283702.htm. Accessed August 2011.

Mullaney, Thomas S. "Ethnic Classification Writ Large: The 1954 Yunnan Province Ethnic Classification Project and its Foundations in Republican-Era Taxonomic Thought." *China Information* 18 (2004): 207–241.

Mullaney, Thomas S. *Coming to Terms with the Nation: Ethnic Classification in Modern China*. Berkeley: University of California Press, 2011.

"Names Unveiled for Panda Pair for Taiwan." *Xinhua*, January 28, 2006. news.xinhuanet.com/english/2006-01/28/content_4113317.htm.

Nanchong shifan xueyuan Wanglang ziran baohu qu da xiongmao diaocha dui (Nanchong Teacher's College Wanglang Nature Reserve giant panda survey team). "Sichuan sheng Pingwu xian Wanglang baohu qu da xiongmao zaihou huifu qingkuang diaocha baogao" (Sichuan province, Pingwu county Wanglang Nature Reserve survey report on the situation following the giant panda crisis). *Nanchong shiyuan xuebao* (Nanchong Teachers College Bulletin) 2 (1984): 41–46.

Naughton, Barry. *The Chinese Economy: Transition and Growth*. Cambridge, MA: Massachusetts Institute of Technology Press, 2007.

"Nei Menggu dapi ru zhipin gongying shichang" (A large supply of Inner Mongolian milk products hit the market). *Renmin ribao* (*People's Daily*), August 24, 1963, 2.

Nemy, Enid. "First Lady's Quest: 'The Real China.'" *New York Times*, April 23, 1984, A6.

Nemy, Enid. "At End of Day for Reagans, Dinner with Nine Courses." *New York Times*, April 28, 1984, 4.

New York Times. 1972, 1984.

Nicholls, Henry. *The Way of the Panda: The Curios History of China's Political Animal*. New York: Pegasus Books, 2011.

Nicholson-Lord, David. "Green Tragedy: The Blight of Ecotourism." *Resurgence*, June 13, 2002 on *AlterNet* http://www.alternet.org/story/13371?page=1. Accessed August 12, 2010.

Nie, Guolin [pseudonym], former surveyor and staff in Wanglang reserve. Interview by author, Pingwu County, Sichuan Province, PRC, October 19, 2001.

"Nikesong zongtong furen zai jing canguan youlan" (President Nixon's wife on a sightseeing tour of the capital). *Renmin ribao* (*People's Daily*), February 23, 1972, 2.

"Nuli guanche zhixing Mao Zhuxi geming wenyi luxian de xin chengguo guoqing qijian zhuang shang ying yi pi xin ying pian" (Diligently carrying out Chairman Mao's line on art for national day with many new films). *Renmin ribao* (*People's Daily*), September 30, 1975, 4.

Oakes, Tim. *Tourism and Modernity in China*. New York: Routledge, 1998.

O'Brien, Stephen J., Pan Wenshi, and Lü Zhi. "Pandas, People and Policy." *Nature* 369 (May 19, 1994): 179–180.

Oreskes, Naomi. "Objectivity or Heroism? On the Invisibility of Women in Science." *Osiris*, Second Series, 11, *Science in the Field* (1996): 87–113.

"Outline of the Protection Project for the Giant Panda and Its Habitat in Sichuan Province." List of Proposals, submitted to WWF, 1993.

Ouyang, Huiyun. "Wolong ziran baohu qu da xiongmao zhuan yi dao anquan didai" (Moving giant pandas in the Wolong Nature Reserve to a safer area). *Renmin ribao* (*People's Daily*), October 22, 1983, 3.

Pan, Wenshi, Lü Zhi, Zhu Xiaojian, Wang Dajun, Wang Hao, Long Yu, Fu Dali, and Zhou Xin. *Jixu shengcun de jihui* (A chance for lasting survival). Beijing: Beijing Daxue chuban she, 2001.

Pan, Wenshi, Gao Zhengsheng, Lü Zhi, Xia Zhengkai, Zhang Miaodi, Ma Cailing, Meng Guangli, Zhe Xiaoye, Lin Xuzhuo, Cai Haiting, and Chen Fengxiang. *Qinling da xiongmao de ziran bihu suo* (The giant panda's natural refuge in the Qinling Mountains). Beijing: Beijing Daxue chuban she, 1988.

"Panda Ambassador Mei Lan Ushers in Chinese New Year by Launching Earth Hour 2010 to the World." *WWF* (World Wide Fund for Nature) *Global*, February 11, 2010, http://wwf.panda.org/wwf_news/?188762/Panda-Ambassador-Mei-Lan-ushers-in-Chinese-New-Year--by-launching-Earth-Hour-2010-to-the-world. Accessed January 21, 2012.

"Panda Attacks Man in Chinese Zoo." *BBC News*. November 22, 2008, http://news.bbc.co.uk/2/hi/7743748.stm. Accessed August 21, 2017.

"Panda-monium in Washington." *New York Times*, June 11, 1972.

"Pandas Extend Coast Visit." *The New York Times*. October 1, 1984, B11.

"Party Draws Ambassadors, Socialites—and Pandas." *The Associated Press*, April 16, 1982.

Peddie, Claire. "Who Do the Pandas Pick?" *Adelaide Now*, July 19, 2010, http://www.adelaidenow.com.au/news/in-depth/who-do-the-pandas-pick/story-fn5rizbk-1225893660105. Accessed July 24, 2010.

"Peking Zoo Willing to Give a Pair of Giant Pandas to Zoo in Taiwan." *Xinhua News Agency*, April 10, 1987.

Peng, Rui, Bo Zeng, Xiuxiang Meng, Bisong Yue, Zhihe Zhang, and Fangdong Zou. "The Complete Mitochondrial Genome and Phylogenetic Analysis of the Giant Panda (*Ailuropoda melanoleuca*)." *Gene* 397 (2007): 76–83.

"Peng, Zhen. Shizhang zenggei Pingrong yi pi zhengui dongwu" (Mayor Peng Zhen gave Pyongyong several precious animals). *Renmin ribao* (*People's Daily*), June 8, 1965, 4.

"The People's Pandas." *New York Times*, March 15, 1972, 46.

Pepper, Suzanne. "Learning for the New Order." In *The Cambridge History of China, Volume 14, The People's Republic, Part I: The Emergence of Revolutionary China, 1949–1965*, ed. Denis Twitchett and John K. Fairbank. Cambridge: Cambridge University Press, 1987.

Pinchot, Gifford. *Breaking New Ground.* Seattle: University of Washington Press, 1947.

PRC, Guowu yuan (State Council). "Guowu yuan dui linye bu guanyu kaizhan wo guo shoulie shiye baogao de bifu" (Report on the State Council address to the ministry of forestry concerning the development of the our national hunting industry) in *Guanyu yesheng dongwu ziyuan de wenjian ji* (Collection of documents concerning wild animal resources). ed. Lin zheng baohu si (Forest protection department). Beijing: Linye bu, 1958.

PRC, Guowu yuan (State Council). "Guowu yuan guanyu jiji baohu he heli liyong yesheng dongwu zi yuan de zhishi" (State Council directive calling for the active protection and rational use of wild animal resources). Guolin Tan zi 287 hao (14 September 1962).

PRC, Guowu yuan (State Council). "Zhonghua renmin gong gong he linye bu wei tingzhi shengchan he bianyong shi yong pu shou wan de lianhe tongzhi" (Announcement to People's government and the Ministry of Forestry calling for the end of the production and sale of specific animal poison) in *Guanyu yesheng dongwu ziyuan de wenjian ji* (Collection of documents concerning wild animal resources), ed. Lin zheng baohu si (Forest protection department). Beijing: Linye bu, 1961.

PRC, Guowu yuan, Linye bu (State Council and Ministry of Forestry), "Guanyu guoying linchang jingying shoulie shiye de ji xiang guiding" (Regulations concerning the operation and management of the hunting industry in national forest areas) in *Guanyu yesheng dongwu ziyuan de wenjian ji* (Collection of documents concerning wild animal resources), ed. Lin zheng baohu si (Forest protection department). Beijing: Linye bu, 1961.

PRC, Linye bu (Ministry of Forestry). "Guanyu guoying linchang jingying guanli shoulie shiye de ji xiang guiding" (Concerning the Operation and Management of Hunting Industry Regulations in National Forest centers). Lin jing Luo zi di 145 hao (10 May 1962).

PRC, Nonglin bu (Ministry of Agriculture and Forestry). "Guanyu jiaqiang da xiongmao baohu gongzuo de jiji tongzhi" (Urgent notice concerning giant panda protection work). Nonglin 76 lin zi di 20 hao (16 March 1976).

PRC, Nonglin bu (Ministry of Agriculture and Forestry). "Guanyu zheng qiu dui 'Yesheng dongwu ziyuan baohu tiaoli' (caoan) de yijian he zanting pizhuo buzu guojia zhengui dongwu de tongzhi" (Report concerning the government's request about the suggestion to temporarily halt the approval to trap precious species with the "Wild animal resource protection regulations' (draft)). 3 Nonglin lin zi di 53 hao (8 May 1973).

Qian, Danning, ed. *Pingwu xianzhi* (Pingwu county gazetteer). Chengdu: Sichuan kexue jishu chuban she, 1997.

Reid, Donald G. and Hu Jinchu. "Giant Panda Selection Between Bashania Fangiana Bamboo Habitats in Wolong Reserve, Sichuan, China." *Journal of Applied Ecology* 28 (1991): 228-243.

Reid, Donald G., Alan H. Taylor, Hu Jinchu, and Qin Zisheng. "Environmental Influences on Bamboo *Bashania fangiana* Growth and Implications for Giant Panda Conservation." *Journal of Applied Ecology* 28 (1991): 855-868.

Reiger, John F. *American Sportsmen and the Origins of Conservation*. New York: Winchester Press, 1975.

Renmin ribao (*People's Daily*), (Abbrev. RMRB), November 28, 1963, 2.

Renmin ribao (*People's Daily*), (Abbrev. RMRB), January 1975–December 1987.

"Report Finds Taiwan Leads in Economic Growth." *Journal of Commerce*, August 22, 1986.

Rigger, Shelley. "Taiwan's Presidential and Legislative Elections." *Orbis* (Fall 2008): 689–700.

Robertson, Nan. "New Pandas melt Hearts at National Zoo." *New York Times*, 18 April 1972, 1.

Roosevelt, Theodore. *The Wilderness Hunter: An Account of the Big Game of the United States and Its Chase with Horse, Hound, and Rifle*. New York: G.P. Putnam's Sons, 1893.

Roosevelt, Theodore and Kermit Roosevelt. *Trailing the Giant Panda*. New York: Blue Ribbon Books, Inc., 1929.

Rothschild, N. Harry. *Wu Zhao: China's Only Woman Emperor*. New York: Pearson Longman, The Library of World Biography, 2008.

Salisbury, Harrison E. "Dinner with Mrs. Sun Yat-sen in Old Peking." *New York Times*, 3 June 1972, 2.

Salisbury, Harrison E. "U.S. Musk Oxen Recuperating, Draw Crowds in Peking Zoo." *New York Times*, 14 June 1972, 2.

San Diego Zoo site: November 8, 2012, http://blogs.sandiegozoo.org/2012/11/06/exam-12-confident-and-curious/#comments. Accessed November 19 and December 3, 2012.

Schaller, George B., Hu Jinchu, Pan Wenshi, and Zhu Jing. *The Giant Pandas of Wolong*. Chicago: Chicago University Press, 1985.

Schaller, George B. *The Last Panda*. Chicago: University of Chicago Press, 1993.

Schmalzer, Sigrid and E. Elena Songster. "Wild Pandas Wild People: Two Views of Wilderness in Deng-Era China." In *Visualizing Modern China: Image, History, and Memory, 1750–Present*, James A. Cook, Joshua Goldstein, Matthew D. Johnson, and Sigrid Schmalzer, eds., 259–278. Lanham, MD: Lexington Books, 2014.

Schneider, Laurence. *Biology and Revolution in Twentieth-Century China*. Lanham: Rowman and Littlefield, 2003.

Schneider, Laurence. "Editor's Introduction: Lysenkoism in China: Proceedings of the 1956 Qingdao Genetics Symposium." *Chinese Law and Government* 19, no 2 (1986): iii–xxi.

Scott, James C. *The Moral Economy of the Peasant: Rebellion and Subsistence in Southeast Asia*. New Haven: Yale University Press, 1977.

"Security is Tight as 2 Pandas Land, Gifts from Peking Housed at Zoo in Washington." *New York Times*, 17 April 1972, 5.

Seidensticker, John, John F. Eisenberg, and Ross Simons. "The Tangjiahe, Wanglang, and Fengtongzhai Giant Panda Reserves and Biological Conservation in the People's Republic of China," *Biological Conservation* 28, no. 3 (1984): 217–251.

Shaanxi sheng, Linye ting (Shaanxi Province, Department of Forestry). "Shaanxi sheng shoulie shiye guanli zanxin banfa" (Shaanxi Province People's Committee announcement concerning the issuing of temporary methods of hunting industry management for Shaanxi Province). Huilin Li zi di 793 hao (26 December 1962).

"Shaoyang zhu yi kai xinhua" (Updating bamboo art in Shaoyang) *Renmin ribao* (*People's Daily*), 23 June 1973, 3.

Shapiro, Judith. *Mao's War Against Nature*. Cambridge, Eng.: Cambridge University Press, 2001.

Shen, F., Zhang, Z., He, W., Yue, B., Zhang, A., Zhang, L., Hou, R., Wang, C. and Watanabe, T., "Microsatellite Variability Reveals the Necessity for Genetic Input from Wild Giant Pandas (***Ailuropoda melanoleuca***) into the Captive Population," *Molecular Ecology* 18 (March 2009): 1061–1070. doi: 10.1111/j.1365-294X.2009.04086.x

Shen, Xu. "Zang yi minzu zoulang yu Yunnan di qiang zuqun (Tibetan-Yi ethnic corridor and the Yunnan Di-Qiang ethnicities)." *South-West China Cultural Studies* 5 (2001): 193–239.

Sichuan Forest Department Representative. Interview by author, Chengdu, Sichuan Province, January 2002.

Sichuan sheng, Linye ting (Sichuan provincial department of forestry). "Guanyu guanche zhixing nonglin bu 'guanyu jiaqiang da xiongmao baohu gongzuo de jiji tongzhi' de tongzhi" (Notice regarding carrying out the policy of the Ministry of Agriculture and Forestry 'Urgent notice regarding strengthening giant panda protection work'). Chuanlin zao 76 di 27 hao (23 March 1976).

Sichuan sheng, Linye ting (Sichuan provincial department of forestry). "Guanyu ziran baohu qu gongzuo de chubu yijian" (Suggestions on the introduction of nature reserve work). Lin jing zi di 010 hao (17 February 1965).

Sichuan sheng Linye ting (Sichuan provincial department of forestry). "Qing shouji da xiongmao ziran siwang qingkuang de jiji tongzhi" (Urgent notice requesting the collection of giant pandas that naturally died). Chuanlin zao 76 di 29 hao (9 April 1976).

Sichuan sheng, Mianyang linye ju (Mianyang prefecture forestry bureau). "Guanyu jiaqiang dui da maoxiong xiankuang guancha he dui jianzhu kaihua siwang hou de huifu qingkuang jinxing diaocha zongjie de tongzhi" (Summary notice regarding the strengthening of the monitoring of the present giant panda situation and the survey of the advancement of post-bamboo flowering and die-off recovery). Dilin jingying di 032 hao (13 March 1976).

Sichuan sheng, Pingwu xian geming weiyuanhui (Pingwu county revolutionary committee). "Guanyu baohu da xiongmao de jiji tongzhi" Urgent notice concerning protecting giant pandas). Pingge fa 76 di 12 hao (10 February 1976).

Sichuan sheng Pingwu xian geming weiyuanhui (Pingwu county revolutionary committee). "Guanyu diaocha da xiongmao qingkuang de jiji tongzhi" (Urgent notice regarding a survey the giant panda situation). Pingge fa 76 di 25 hao (13 April 1976).

Sichuan sheng, Pingwu xian geming weiyuanhui (Pingwu county revolutionary committee). "Guanyu zhuan fa Sichuan sheng geming weiyuanhui yingfa nonglin bu song 'guanyu da xiongmao daliang siwang qingkuang de diaocha baogao' de han" (Reply regarding the forwarding of Sichuan province's Revolutionary Committee's issuing of the ministry of agriculture and forestry's directive, 'concerning the survey report of the large number of giant panda deaths'). Pingge fa 76 di 78 hao (14 December 1976).

Sichuan sheng, Pingwu xian geming weiyuanhui (Pingwu county revolutionary committee). "Guanyu zuo hao baohu da xiongmao gongzuo de jiji tongzhi" (Urgent notice concerning doing giant panda protection work well). Pingge fa 76 di 29 hao (22 April 1976).

Sichuan sheng, Pingwu xian, Linye ju (Pingwu county forestry bureau). "Guanyu caoni 'yesheng dongwu ziyuan baohu tiaojian' de shuoming" (Draft of explanation of 'factors in terms of wild animal resource protection'). (May 1973).

Sichuan sheng, Pingwu xian, Linye ju (Pingwu county forestry bureau). "Guanyu jiaqiang Wanglang ziran baohu qu guanli gongzuo de baogao" (Report concerning the strengthening of Wanglang Nature Reserve management work). Pinglin hu 76 zi di 03 hao (20 May 1976).

Sichuan sheng, Pingwu xian, Linye ju (Pingwu county forestry bureau). "Guanyu qingshi jiejue si da xiongmao shiliang gongying de baogao" (Report concerning request to solve the problem with supplies for feeding and raising giant pandas). Pinglin ban 76 zi di 06 hao (22 June 1976).

Sichuan sheng, Pingwu xian, Linye ju (Pingwu county forestry bureau). "Guanyu Wanglang ziran baohu qu 76 nian jijian kuan kai ji jihua baogao" (Report concerning Wanglang Nature Reserve construction budget for 1976). Pinglin ban76 zidi 04 hao (18 June 1976).

Sichuan sheng, Pingwu xian, Linye ju (Pingwu county forestry bureau). "Pingwu xian shoulie gongzuo kaizhan qingkuang jianjie" (A concise report on the status of the development of work on hunting in Pingwu county) (Sept. 13, 1965).

Sichuan sheng, Pingwu xian, Renmin weiyuanhui (Pingwu county people's committee). "Guanyu jianli Wanglang ziran baohu qu de baogao" (Concerning the establishment of Wanglang Nature Reserve). Hui ban zi di 022 hao. (1 July 1965).

Sichuan sheng, Renmin weiyuan hui pizhuan sheng linye ting (Sichuan Provincial people's committee transmission to the provincial department of forestry). "Guanyu jiji baohu he heli liyong yesheng dongwu ziyuan de baogao" (Report concerning the active protection and rational use of wild animal resources). Chuan nong zi 0191 hao (2 April 1963).

Sichuan sheng, Renmin weiyuan hui (Sichuan provincial people's committee). "Guanyu tongyi jianli Wanglang ziran baohuqu de pifu, (Reply concerning the approval of creating the Wanglang Nature Reserve). Chuan lin zi di 0511 hao (20 September 1965).

Sichuan sheng, Yesheng dongwu ziyuan diaocha baohu guanli zhan he Sichuan sheng linye kexue yanjiu yuan (Sichuan province wild animal resources survey and protection management station and Sichuan Provincial forestry science research institute). *Sichuan Wanglang baohu qu zonghe kexue kaocha baogao* (Sichuan, Wanglang nature reserve integrated scientific investigation report). Chengdu: Sichuan sheng yesheng dongwu ziyuan diaocha baohu guanli zhan he Sichuan sheng linye kexue yanjiu yuan, 1999.

"Sichuan Wanglang, Tangjiahe ziran baohuqu xuexi kaocha baogao (Sichuan, Wanglang, Tangjiahe nature reserve study and survey report)." Mianyang shi linyeju zhuban (Mianyang City Forestry Office), December, 30, 2005.

Sichuan zhuan (Sichuan volume). Zhongguo ziran ziyuan cong shu (Chinese natural resources series), ed. Fu Yingquan, no. 33. Beijing: Zhongguo huanjing kexue chuban she, 1995.

Siku quanshu (Complete books of the four treasuries). Wen yuan ge si ku quan shu dianzi ban (electronic source). Hong Kong: Dizhi wenhua chuban youxian gongsi, Zhongwen daxue chuban she, 1999.

Smil, Vaclav. *China's Environmental Crisis.* Armonk, NY: M.E. Sharpe, 1993.

Sobel, Richard, William-Arthur Haynes, and Yu Zheng. "The Polls—Trends: Taiwan Public Opinion Trends, 1992-2008: Exploring Attitudes on Cross-Strait Issues." *Public Opinion Quarterly* 74, No. 4 (Winter 2010): 782–813.

Song, Houqing. "Linye bu fu buzhang Dong Zhiyong shuo: qiangjiu da xiongmao shi xiang chang qi gongzuo" (The vice-chair of the Ministry of Forestry, Dong Zhiyong said: saving the giant panda is a long-term task). *Renmin ribao* (*People's Daily*), 18 December 1984, 3.

Song, Lianfeng. "Wolong leng jian zhu mianji kaihua gusi da xiongmao wunai xia shan jieshi huagu zhu" (Because the arrow bamboo has mass-flowered and died in Wolong, giant pandas have no choice but to descend to eat the umbrella bamboo). *Renmin ribao* (*People's Daily*), 22 April 1984, 1.

Spence, Jonathan D. *The Search for Modern China*. New York: W.W. Norton & Co., 2013 [1990].

Steinburgh, James. "Hua Mei's Trip to China Postponed Again, This Time by SARS Fears; Human Companions Would Be Put at Risk." *Union Tribune* (San Diego), April 23, 2003.

Strapagiel, Lauren. "Nishiyuu Journey By Cree Youth Ends As Harper Greets Pandas." *Huffington Post*, March 25, 2013, http://www.huffingtonpost.ca/2013/03/25/nishiyuu-journey-ends-ottawa-harper-pandas_n_2950643.html (Accessed July 27, 2017); Emily Mertz, "Harper's Panda Meeting Sparks Criticism," Global News, March 25, 2013, http://globalnews.ca/news/427585/harpers-panda-meeting-sparks-criticism/. Accessed August 21, 2017.

Stronza, Amanda. "Anthropology of Tourism: Forging New Ground for Ecotourism and Other Alternatives." *Annual Review of Anthropology* 30 (2001): 261–283.

Sun, Nongzhai, ed. *Sichuan sheng ditu ce*. Chengdu: Chengdu ditu chuban she. 2000.

Suttmeier, Richard P. "Party Views of Science: The Record from the First Decade." *The China Quarterly*, 44 (Oct.–Dec., 1970), 146–168.

Suttmeier, Richard P. *Research and Revolution: Science Policy and Societal Change in China*. Lexington, MA: Lexington Books, 1974.

Swaisgood, Ronald R., Zejun Zhang, Fuwen Wei, David E. Wildt, and Andrew J. Kouba. "Giant Panda Conservation Science: How Far We Have Come." *Biology Letters* 6.2 (2010): 143–145.

"Taipei Zoo Challenges COA's Panda Rejections." *China Post*, June 25, 2007.

"Taiwan Rejects China's Giant Pandas." *IOL*, March 31, 2006 http://www.iol.co.za/general/news/newsprint.php?art_id=qw1114380. Accessed February 27, 2008.

"Taiwan says 'No' to China's Panda Offer." *Straits Times*, July 24, 1990.

"Taiwan Turns Down Mainland Panda Offer." *Japan Economic Newswire*, December 9, 1988.

"Tanaka Shouxiang huijian Liao Chengzhe tuanzhang Riben xinwen jie pengyou juxing cha hui huanying wo daobiao tuan" (Prime Minister Tanaka Kakuei met committee head Liao Chengzhi Japan's news friends hosted a tea ceremony to welcome our representative committee). *Renmin ribao* (*People's Daily*), 15 May 1973, 5.

Tang, Tansheng. "Chuan bei jiu xian de maopi shou ji qi liyong de chubu baogao" (Report on the early steps in finding pelt animals and other useful items in the northern nine counties of Sichuan). *Dongwuxue zazhi* (Journal of zoology) 4, no. 5 (1960): 195–199.

Taylor, Alan H. and Qin Zisheng. "Culm Dynamics and Dry Matter Production of Bamboos in the Wolong and Tangjiahe Giant Panda Reserves, Sichuan, China." *Journal of Applied Ecology* 24, no. 2 (August 1987): 419–433.

Taylor, Alan H., Qin Zisheng, Liu Jie, "Structure and Dynamics of Subalpine Forests in the Wang Lang Natural Reserve, Sichuan, China," *Vegetation* 124 (1996): 24–38.

TIES, The International Ecotourism Society, "TIES Global Ecotourism Factsheet," TIES, September, 2006, www.ecotourism.org. Accessed July 7, 2011.

Tsuchiya, Masaya. "Recent Developments in Sino-Japanese Trade." *Law and Contemporary Problems* 38, no. 2, Trade with China (Summer-Autumn, 1973): 240–248.

"Two Pandas are Presented to Bronx Zoo by Chinese at Ceremony in Chungking." *New York Times*, November 10, 1941, 19.

"Two Pandas Selected to Move to Macau," *Macau Daily Times* online, May 25, 2010. http://www.macaudailytimes.com.mo/. Accessed May 25, 2010.

Union Tribune (San Diego). December 2, 4, 1999.

United Nations. Resolution 2758 (XXVI). "Restoration of the lawful rights of the People's Republic of China in the United Nations." 1976th plenary meeting, October 25, 1971. http://daccess-dds-ny.un.org/doc/RESOLUTION/GEN/NR0/327/74/IMG/NR032774.pdf?OpenElement. Accessed July 31, 2010.

Wallace, D. R. *The Quetzal and the Macaw*. San Francisco: Sierra Club Books, 1992.

Wallace, Gregory. "'Panda Cam' Goes Dark in Shutdown." *CNN, Money*, October 1, 2013. http://money.cnn.com/2013/09/30/news/panda-cam-national-zoo/index.html. Accessed August 21, 2017.

Wang, Dajun. "Postscript: The People of the Qinling Study." In Pan Wenshi, Lü Zhi, Zhu Xiaojian, Wang Dajun, Wang Hao, Long Yu Fu Dali, and Zhou Xin, *A Chance for Lasting Survival: Ecology and Behavior of Wild Giant Pandas*, trans. By Richard B. Harris. Washington, D.C.: Smithsonian Institution Scholarly Press, 2014.

Wang, Shengyan. "Yi minzu maoyi gongzuo zai Baima zang qu de kaichuang (Remembering the initiation of trade work with the minorities in the Baima Tibetan area)." In *Pingwu wenshi ziliao xuanji* (Collection of personal histories from Pingwu County). Pingwu: Pingwu caiying chang yingshuang, 2000.

Wang, Zuoye. "Zhu Kezhen." In *Complete Dictionary of Scientific Biography* ed. Noretta Koertge. New York: Charles Scribner's Sons, 2007.

Wanglang National Nature Reserve website: http://www.wanglang.com/Display.asp?ID=26 Accessed August 12, 2010.

Wanglang Nature Reserve TEAM Network webpage, http://www.teamnetwork.org/en/field_stations/wanglang-nature-reserve. Accessed August 7, 2010.

Wanglang ziran baohu qu da xiongmao diaocha zu (Wanglang nature reserve giant panda survey team). "Sichuan sheng Pingwu xian Wanglang ziran baohu qu da xiongmao de chubu diaocha (Preliminary survey of giant pandas in the Wanglang nature reserve in Pingwu County, Sichuan). *Dongwuxue bao* (Acta Zoologica Sinica) 20, no. 2 (June 1974): 162–173.

Watts, John. "1,3000 Years of Global Diplomacy Ends for China's Giant Pandas." *The Guardian International*, September 14, 2007, 27.

Weaver, David B. "Magnitude of Ecotourism in Costa Rica and Kenya." *Annals of Tourism Research* 26, no. 4 (1999): 792–816.

Weiner, Douglas R. *Models of Nature: Ecology, Conservation, and Cultural Revolution in Soviet Russia*. Pittsburgh: University of Pittsburgh Press, 1988.

Weiner, Douglas R. *A Little Corner of Freedom: Russian Nature Protection from Stalin to Gorbachëv*. Berkeley: University of California Press, 1999.

Wen, Zhe and Wang Menghu. "Da xiongmao yu zhu" (Giant pandas and bamboo). *Da ziran* (Nature) 1 (1980): 12–14.

"Wenhua da geming tuidong zhaoxiangji gongye da fazhan: 1976 nian quanguo zhaoxiangji de chanliang, bi Wenhua Da Geming qian de 1976 nian zengchangle 12 bei, pinzhong zengjia jin 4 bei. Shishi you li de pipan le Deng Xiaoping 'Jin buru xi'de miulun'" (The Great Proletarian Cultural Revolution promotes the development of the camera industry: In 1976 the production of cameras throughout the entire country was greater than before the Great Proletarian Cultural Revolution by twelve fold, the number of camera models increased by four fold offering a genuine criticism of Deng Xiaoping's false assertion that 'Today is not as good as in the past'). *Renmin ribao* (*People's Daily*), 14 June 1976, 3.

Wilson, E.H. *A Naturalist in Western China*. London: Cadogan Books, 1913.

"Wo dui wai xie daibiao tuan jiesu dui Yilang youhao fangwen" (Our foreign relations envoy goes to Iran). *Renmin ribao* (*People's Daily*), 11 September 1977, 6.

"Wo dui wai you xie daibiao tuan fangwen Yilake" (Our foreign relations envoy goes to Iraq). *Renmin ribao* (*People's Daily*), 27 September 1977, 5.

"Wo guo xinxing jiaopian gongye dale fanshen zhang" (Our nation brought about an upswing in the burgeoning film industry). *Renmin ribao* (*People's Daily*), 12 February 1976, 1.

"Wo guo zhengfu zengsong Pengpishe zongtong he Faguo renmin de yi dui da xiongmao yunwang Pali" (The pair of pandas that our national government gave President Pompidou and the people of France are headed for Paris). *Renmin ribao* (*People's Daily*), December 9, 1973, 4.

"Wo keji yingpian zai Luoma fangying shoudao huanying" (Our technology film was well received in Rome). *Renmin ribao* (*People's Daily*), 17 December 1976, 6.

"Wo yingpian zai Yindu guoji dianying jie shoudao huanying" (Our films were well received in India's international film festival). *Renmin ribao* (*People's Daily*), 2 February 1977, 5.

"Wo zengsong de yidui daxiongmao yun di Moxige" (Our gift pandas have arrived in Mexico), *Renmin ribao* (*People's Daily*), September 12, 1975, 5.

"Wo zengsong xiongmao jiaojie yishi zai pali juxing Ceng Tao dashi he Fengdanei bu zhang gong zhu Zhong Fa youyi bu duan fazhan" (At the ceremony for our gift of pandas, Ambassador Ceng Tao and Ministry Head Fengdanei together celebrated the continuous development of friendship between China and France), *Renmin ribao* (*People's Daily*), December 22, 1973, 5.

"Wo zengsong Yingguo renmin da xiongmao jioajie yishi zai lundun juxing" (The handover ceremony for the giant pandas we gave to the English people was held in London). *Renmin ribao* (*People's Daily*), 8 November 1974, 6.

"Woman Accused of Peddling Panda Pelt." *USA Today*, July 23, 2008.

World Wide Fund for Nature (WWF), China. http://www.wwfchina.org/english/loca.php?loca=107#2. Accessed December 18, 2012.

World Wide Fund for Nature (WWF), Earth Hour Website: http://www.andymurray.com/news/andy-murray-serves-up-earth-hour-challenge-to-the-uk/; http://www.wwf.eu/?207918/actress-jessica-alba-announced-as-earth-hour-2013-global-ambassador. Accessed on July 24, 2015.

Wren, Christopher S. "Bureaucracy and Blight Imperil China's Pandas." *New York Times*, 3 July 1984, C1.

Wren, Christopher S. "Chinese Official Denies Gift for Pandas was Sidetracked," *New York Times*, 17 October 1984, A19.

Wu, Lilian. "Pandas May Come to Taiwan." *Central News Agency* (Taiwan), June 6, 1997.

Wu, Zhonglun, ed., *Zhongguo ziran baohu qu* (China's nature reserves). Shanghai: Keji jiaoyu chuban she, 1996.

WuDunn, Sheryl, "Pessimism Is Growing on Saving Pandas From Extinction." *New York Times*, June 11, 1991, C4

Xi, Zefu. "Chengdu shi di shi zhongxue shi sheng jiji juankuan zhengjiu da xiongmao" (Teachers and students and Chengdu's Number Ten middle school enthusiastically contributed funds to save the giant panda). *Sichuan dongwu* (Sichuan Animals) 3, no. 3 (August 1984): 46.

Xiao, Youyuan. *Pingwu Baima Zang zu* (The Baima Tibetans in Pingwu County). Mianyang: Mianyang hong guang qiye yingshua chang, 2001.

Xie, Zhong and Jonathan Gipps. *The 2001 International Studbook for Giant Panda* [sic] *(Ailuropoda melanoleuca)*. Beijing and London: Chinese Association of Zoological Gardens and Bristol Zoo Gardens, 2001.

"Xin, Xilan zongli Maerdeng fangwen Qinghua Daxue" (New Zealand Prime Minister Martin visits Qinghua University). *Renmin ribao* (*People's Daily*), 30 April 1976, 2.

Xu, Dixin ed. *China Conservation Strategy*. Beijing: United Nations Environment Programme and China Environmental Science Press, 1990.

Yan, Jiaqi and Gao Gao. *Turbulent Decade: A History of the Cultural Revolution*. trans. D.W.Y. Kwok. Honolulu: University of Hawai'I Press, 1996.

Yang, Li, ed. "Taiwan Rejects Pandas." *China Daily*, April 1, 2006, www.chinaview.cn 2006-04-01 09:37:05. Accessed July 21, 2010.

Yang, Meng-yu, "Taiwan wucheng minzhong huanying xiongmao daolai, [In Taiwan fifty percent of the populace welcome the arrival of pandas]" *Lienhe bao*, January 16, 2006.

Yang, Meng-yu. "Xiongmao mei lai; kao ya lai." *BBC Chinese.com*, July 6, 2007, http://newsvote.bbc.co.uk/mpapps/pagetools/pring/news.bbc.co.uk/ch. Accessed on July 22, 2010.

Yang, Ruoli, Zhang Fuyun, and Luo Wenying. "1976 nian da xiongmao zainan xing siwang yuanyin de shenlun" (Probing into reasons for the 1976 catastrophic death of giant pandas). *Acta Theriologica Sinica* 1, no. 2 (December 1981): 127-134.

Yang, Su, "Mass Killings in the Cultural Revolution: A Study of Three Provinces," in *The Chinese Cultural Revolution as History* eds. Joseph W. Esherick, Paul G. Pickowicz, and Andrew Walder, 96-123. Stanford: Stanford University Press, 2006.

Yao, Shuping. "Chinese Intellectuals and Science: A History of the Chinese Academy of Sciences (CAS)." *Science in Context* 3 no. 2 (1989):447-473.

Yao, Weimin, ed. *Zhonggong Pingwu difang shi dashi ji, 1935-1998* (Central Pingwu local history and record of major events, 1935-1998). Pingwu: Sichuan donghua yingwu jituan youxian gongsi, 1999.

Yaussy, Nathan. Creator of EUT, Endangered Ugly Things, website, posted purpose of website. http://endangered-ugly.blogspot.com/. Accessed December 18, 2010.

Yeh, Emily and Chris Coggins, eds., *Mapping Shangrila: Contested Landscapes in the Sino-Tibetan Borderlands.* Seattle: University of Washington Press, 2014.

Yin, Hong. "Zhongguo yesheng dongwu baohu xiehui fuze ren tan muqian da xiongmao de zaiqing ji jixu de jiuzai cuoshi (Head of China's Wild Animal Protection Association discusses the present situation of the giant panda crisis and the urgent need for panda rescue measures)." *Renmin ribao* (*People's Daily*), 16 April 1984, 3.

"Ying hua shi jie fang dong ling" (Visiting eastern neighbors during the cherry blossom festival). *Renmin ribao* (*People's Daily*), 10 May 1973, 5.

"Youguan da xiongmao de lishi jizai" (A written account of the history of the giant panda). *Zhongguo ribao wangzhan* (*China Daily Online*), China Culture.org http://www.chinaculture.org/gb/cn_zggd/2006-01/11/content_77854.htm.

"Youqu de suliao wanju" (An interesting plastic toy). *Renmin ribao* (*People's Daily*), 29 January 1962, 2.

Yu, Sophie. "Taiwan to Accept Pandas." *The Times* (London), March 24, 2008.

Zeng, Weiyi. "Baima ren zuyu yanjiu jian jie," (A simple introduction of research on the classification of the Baima people). In *Baima zangzu yanjiu wenji*, (Collected work on research on the Baima Tibetans). Edited by Zeng Weiyi. Chengdu: Sichuan sheng minzu yanjiu suo, 2002.

Zhan, Xiangjiang, Ming Li, Zhejun Zhang, Benoit Goossens, Youping Chen. "Molecular Censusing Doubles Giant Panda Population Estimate in a Key Nature Reserve." *Current Biology* 16.12 (2006): 451-452.

Zhang, Ding and Zhang Mei. "Shiyong meishu—Zhongying gongyi meishu xueyuan qizuo shixi" (Useful art—Work and practice of the Central Craft and Art Academy). *Renmin ribao* (*People's Daily*), 4 June 1961, 8.

Zhang, Shougong. "China's 1998 Flood Disaster, Cause and Response." Presented at "Natural Disaster and Policy Response in Asia, Implications for Food Security Conference." Harvard University Asia Center, Spring 1999.

Zhang, Xiruo daibiao de fayan (Representative Zhang Xiruo's speech). "Huxiang xiqu, huxiang fazhan, huxiang zunzhong fazhan dui wai wenhua jiaoliu gongzuo" (Mutual assimilation, mutual development, mutual respect, development of cultural foreign exchange work). published in *Renmin ribao* (*People's Daily*), 4 May 1959, 5.

Zhang, Zhenguo et al. "Ecotourism and Nature-Reserve Sustainability in Environmentally Fragile Poor Areas: The Case of the Ordos Relict Gull Reserve in China." *Sustainability: Science, Practice, & Policy* 4, No. 2 (Fall/Winter 2008), 13–14. http://ejournalnbii.org. Accessed August 8, 2010.

Zheng, Zuoxin. "Sulian dongwu qu xi yu shoulie yanjiu shiye de fazhan jinkuang" (Recent status of the development of Soviet animal reserves and hunting research industries). *Dongwuxue zazhi* [Journal of zoology] 3, no. 1 (1959): 31-33.

"Zhong-Ri youhao de xin bianzhang—ji Zhong-Ri youhao xiehui daibiao tuan fangwen Riben" (New chapters in Chinese-Japanese friendship—records of the Chinese-Japanese friendship association representatives' visit to Japan). *Renmin ribao* (*People's Daily*), 4 June 1973, 5.

Zhong, Zhaomin. "Da xiongmao siwang yu dizhen de gaunxi" (The relationship between giant panda death and earthquakes). Unpublished, undated essay.

Zhong, Zhaomin. "Wanglang zai qianjin" (Wanglang forging forward), unpublished papers, April 2004.

Zhong, Zhaomin. "Zai Wanglang ziran baohu qu paishe kejiao yingpian, 'Xiongmao' gushipian, 'Xiongmao lixian yanzhi' deng yingpian de gaishu [Overview of the filming of the science education documentary, 'Panda' and the feature film 'Diary of a panda's adventures' in the Wanglang Nature Reserve]." Unpublished personal reflections, April 2004.

Zhong, Zhaomin. "Zhulei kaihua yu dizhen de guanxi," undated essay, 2.

"Zhongguo chuji zhi sheng de xionmao zhengzhi" [China's panda politics subdues opposition], *Duoweiboge* blog, April 5, 2006. Accessed April 5, 2006.

Zhongguo xibei diqu "Zai Wanglang ziran baohu qu paishe kejiao yingpian, 'Xiongmao' gushipian, 'Xiongmao lixian yanzhi' deng yingpian de gaishu [Overview of the filming of the science education documentary, 'Panda' and the feature film 'Diary of a panda's adventures' in the Wanglang Nature Reserve]," personal reflections, April 2004. *Zhenxi binwei dongwu zhi* (Precious and rare animals of northwestern China). Beijing: Zhongguo linye chuban she, 1994.

Zhongguo youpiao quanji Zhongguo Renmin Gongheguo juan. Beijing: Beijing Yanshan chuban she, 1989.

Zhou, Jianren. "Guanyu xiongmao" (On giant pandas). *Renmin ribao* (*People's Daily*), (6 July 1956), 8.

Zhu, Jing. "Dui shoulie qu xi gongzuo ji shoulie qu xi diaocha fa de zu qian renshi (xia)" (Preliminary findings from a survey of the animal reserve system). *Dongwuxue zazhi* (Journal of zoology) 3, no. 8 (1959): 384-385.

Zhu, Jing. "Jiefang yihou wo guo de shoulie, xunyang, yu baohu gong zuo" (National hunting, rearing, and nature protection work since liberation). *Dongwuxue zazhi* (Journal of zoology) 6, no. 6 (1964): 315-316.

Zhu, Jing and Long Zhi, "Da xiongmao de xingshuai (The rise and fall of the giant panda)," *Dongwu xuebao* (*Acta Zoologica Sinica*) 29.1 (March 1983): 93-103.
Zhu, Jing and Li Yangwen. *Da xiongmao* (Giant panda). Beijing: Science Press, 1980.
Zhu, Kezhen. "Kaizhan ziran baohu gongzuo" (Developing nature protection work). Speech delivered at 1963 large meeting of national representatives. In *Zhu Kezhen wen ji* (Collected essays of Zhu Kezhen). Beijing: Kexue chuban she, 1979.
Zhu, Kezhen, "Zhongguo ziran qu hua (chugao) 'xu,' 1959" (Preface of China's nature reserve plan, draft, 1959). In *Zhu Kezhen Wen ji* (Collected essays of Zhu Kezhen. Beijing: Kexue chuban she, 1979.
"Zoo officials withdraw London's aging panda." *New York Times*, 23 April 1972, 10.
"Zoo Sticks to its Guns over Accepting China's Pandas," *China Post*, March 28, 2008.

Index

CPSIA information can be obtained
at www.ICGtesting.com
Printed in the USA
BVHW070253180520
579766BV00004B/12